Topics in Recreational Mathematics 3/2015

Editor-in-chief

Charles Ashbacher
5530 Kacena Ave
Marion, IA 52302 USA

cashbacher@yahoo.com

Assistant Editors

Rachel Pollari

Jennifer Corrigan

Artwork

Caytie Ribble

Problem Contributor

Lamarr Widmer

ISBN: 978-1511641005

Contents

INTRODUCTION

Welcome to the third book in the "Topics in Recreational Mathematics" (TRM) series. Feedback has been very positive, I continue to receive submissions of papers and so am pleased to announce that there will be a fourth book.

As of this book, the alphametics material that was slated to appear in "Journal of Recreational Mathematics" has been essentially exhausted. However, alphametics will continue to be a regular feature of this series and I will be serving as the editor. I encourage readers to submit alphametics to me via one of my addresses listed on the title page.

This book contains a wide variety of papers, something that I am very happy about. I am continuing to conduct a deep dive of the material that I have on file, so there are items that may suddenly be published after years of being packed away.

Other books in recreational mathematics are currently in the development stage. Two where the content is in the process of being created are a book of mathematical cartoons drawn by Caytie Ribble as well as a book of alphametics based on themes from Star Trek©.

As is always the case, I welcome input from readers regarding their opinion of these books.

Charles Ashbacher

cashbacher@yahoo.com

MATHEMATICAL CARTOONS

drawn by Caytie Ribble

ANTECEDENT

$$\int f(x)dx = -C + F(x)$$

CATENARY CURVE

OPERAND

$$\int \frac{d \; ical}{ical} = \log \; ical$$

THE *REAL* CURSE OF THE BAMBINO

Christopher J. Brown

Lyon Carter III

Paul M. Sommers

Department of Economics

Middlebury College

Middlebury, Vermont 05753

psommers@middlebury.edu

Abstract

If Babe Ruth had not spent his first five-plus years as a pitcher with the Boston Red Sox, would he still be wearing the home-run crown? The authors show that if Ruth had been employed as an everyday player early in his career, the probability is greater than 50 percent that he would have hit 900 career home runs. While skeptics might argue that no hitter – not even Babe Ruth – could hit more than 40 home runs per season during the dead ball era (before 1919) to reach the 900 home run mark, the authors show that the Bambino could have exceeded Hank Aaron's 755 career home runs (later eclipsed by Barry Bonds' 762) with near certainty.

I had to wonder if I would ever get out

of that man's shadow.

— Hank Aaron

I Had A Hammer [1]

Introduction

George Herman "Babe" Ruth was arguably the greatest pure slugger of his or any era, although he does not rank first in home runs (second), runs (tied for third), extra base hits (third), or runs batted in (second).[1] Ruth's career began in 1914 as a left-handed pitcher for the Boston Red Sox. In six years on the mound at Fenway, he won 89 regular season games and three World Series games (1-0 in 1916 and 2-0 in 1918). Before he was converted to a full-time outfielder, Ruth had fewer than 150 at bats in each of his first four years with the Red Sox. Apparently, Boston did not begin to recognize his value as a hitter until 1918 when he had 317 appearances. In 1919, Babe's last year in Boston before that now infamous sale to the New York Yankees, he had 432 at bats and led the majors in slugging average (.657), home runs (29),[2] runs (103), and runs batted in (114) [3]. In his first season in Yankee pinstripes, Babe Ruth hit 54 home runs, shattering his own single-season record established the year before.[3] George Sisler was a distant runner-up in 1920 with only 19 home runs.

Baseball fans might now wistfully wonder how much higher Babe Ruth's historic standing would have been if he had had more opportunities to swing his 42-ounce bat earlier in his career. Did those five years as a pitcher cost him a realistic chance to establish and keep multiple all-time batting records, including a reasonable shot at 900 career home runs?

The Data and Results

Table 1 summarizes Babe Ruth's career totals in four different batting categories through sixteen full seasons (excluding his first five as a pitcher for the Red Sox and his final season in 1935 with the Boston Braves for whom he played only 28 games). Babe Ruth's career statistics are from *The Baseball Encyclopedia* [3].

Between 1919 and 1934, Babe Ruth hit 688 home runs ($\mu = 43.0$ home runs per season, $\sigma = 11.12$). To jump ahead of Aaron, Ruth would only need an additional 68

($= 756 - 688$) home runs over the five seasons he spent pitching. Hence, the statistical question becomes: What is the probability that Ruth's five-season average $\bar{x}$ would be at least as large as 13.6 (= 68/5) if his average (as a non-pitcher) was 43.0 home runs per season with a standard deviation of 11.12? Ruth's averages in all four batting categories over sixteen "full" seasons were much higher than the corresponding five-season averages Babe would need to move to the top of the record lists. As can be seen from the uniformly high probabilities reported in the last column of Table 1, if Ruth had been employed differently early in his career he would most likely *still* be *first* in "Home Runs", *first* in "Runs", *first* in "Extra Base Hits", and *first* in "Runs Batted In".

The ease Ruth would have had at displacing Aaron begs the following question: Could the Babe have hit 900 home runs? Now, the statistical question becomes: What is the probability that Ruth's five-season average $\bar{x}$ would be at least 42.4 [= (900-688)/5] if $\mu= 43$ and $\sigma = 11.12$? Clearly, this probability will be greater than 0.5. How much greater? First, we standardize the value $\bar{x} = 42.4$:

$$z = \frac{\bar{x} - \mu}{\sigma / \sqrt{n}} = \frac{42.4 - 43.0}{11.12 / \sqrt{5}} = -.12$$

Thus,

$$\Pr(\bar{x} \geq 42.4) = \Pr(z \geq -.12) = .55$$

Loosely speaking (assuming that the sampling distribution of $\bar{x}$ is roughly normal with mean μ and a standard error of $\sigma / \sqrt{n}$), the probability is .55 that Babe Ruth would have hit 900 career home runs. Had Ruth not pitched on a regular basis early in his career, he would still be wearing the home run crown.

Concluding Remarks

Babe Ruth won 94 games in his career as a pitcher (89 with the Red Sox, 5 more with the Yankees). He won a lot more as a hitter. Admittedly, baseball lost a very good pitcher when Babe Ruth became the Sultan of Swat. Yet, if Ruth had not spent the first five years of his career as a pitcher for the Boston Red Sox, he could have easily surpassed Hank Aaron's career marks. Babe Ruth would today still be the all-time leader in home runs, runs, extra base hits, and runs batted in. Perhaps the real "Curse of the Bambino" can be summed up in a single word: pitching.

Table 1. Babe Ruth's Statistics, 1919 – 1934

Category	Major League Record	Ruth's 16-Season Total	Ruth's 16-Season Average	5-Season Average Needed To Break Record	z_{Calc}	$p(z > z_{Calc})$
Home Runs	755	688	43.00	13.60	-5.72	>.999
Runs	2288	2062	128.875	45.40	-5.91	>.999
Extra Base Hits	1477	1264	79.00	42.80	-3.87	>.999
Runs Batted In	2297	2084	130.25	42.80	-6.67	>.999

Note: Through the 2002 regular season, Rickey Henderson held the record in Runs; Hank Aaron was the all-time major league leader in Home Runs, Extra Base Hits, and Runs Batted In [2].

References

1. H. Aaron, *I Had a Hammer: The Hank Aaron Story*. HarperCollins, New York, 1991.

2. G. Brown and M. Morrison (eds.), *2003 ESPN Information Please Sports Almanac*. Boston, MA, 2002.

3. *The Baseball Encyclopedia* (ninth edition). Macmillan Publishing Company, New York, 1993.

Footnotes

1. At the time of his retirement in 1935, Ruth led in all four categories but one. Ty Cobb (1905-1928) scored 2246 runs, 72 more than the Babe. Hank Aaron (1954-1976) holds the major league records for career extra base hits and runs batted in, and is second only to Barry Bonds for home runs.

2. Babe's 29 home runs in 1919 broke Ned Williamson's single-season home run record of 27 set in 1884.

3. Babe Ruth's 54 home runs in 1920 would have eclipsed the Red Sox' single-season home run record of 50 set by Jimmie Foxx in 1938.

IF MIDY DOESN'T WORK, WHAT THEN?

David L. Emory

137 Sycamore Lane

Lexington, VA 24450

emory@kalexres.kendal.org

Abstract

The research reported here can be compared and contrasted with that of Midy, the French mathematician of the early 1800's. Both involve the calculation of period lengths and adding the first and second halves of the resulting repetends. Midy worked with primes, but here the numbers are composites. One set of factors is 7 and 13, and the other set is 73, 17, and 97. The interesting results here are quite different from Midy's string of 9's.

Introduction

There is something truly magical about the digit 9 [1]. A particularly fascinating example is seen in Midy's Theorem, discovered in 1836 [2]. In considering prime numbers whose reciprocals have even period lengths, Midy found that each even (2N) repetend (the period of a repeating decimal) can be divided into two halves, which when added together produce a string of nines ($10^N - 1$) [3]. For example, when one is divided by seventeen, a repeating decimal is created with a period length of sixteen. The repetend is: 0588235294117647. Adding the two halves gives us:

$$
\begin{array}{r}
05882352 \\
+94117647 \\
\hline
99999999
\end{array}
$$

The proof of Midy's Theorem can be found in an article by Lewittes [4]. Noteworthy here is Lewittes' demonstration that the composite (not prime) number, 803—despite having an even period length—does **not** obey Midy's Theorem.

Additional Exceptions

Reported herein are eight more composite numbers that are, like 803, exceptions to the rule that even period lengths inevitably lead to a string of N nines. Unexpectedly, the nines still show up—but in differing patterns!

Let's do what one would do if the Theorem **were** working: divide the repetend into two halves and add them together. The results are surprising (Tables 1 and 2).

The period lengths of 24, 48 and 96 show, in the last column, how the Midy Theorem treats the 6-digit half-repetend sums in the third column. The two halves added yield three nines. But in each case there are more identical six-digit sums that would also create three nines. Count the identical 6-digit sums, multiply by three (the number of nines produced by Midy, in the last column), and the product will be one quarter of the period length.

Table 1

Adding Two Digits

DIVISOR OF ONE	PERIOD LENGTH	REPETEND	REPETEND HALVES
803 = 11 × 73*	8	00124533	0012 + 4533 4545** 9 9
187 = 11 × 17	16	0053475935828877	00534759 + 35828877 36363636† 9 9 9 9

* Reference 4.　　　**Add 4+5 twice.　　　†Add 3+6 four times.

Analysis and Conclusions

(1) When Midy works, the number of nines is half of the period length. Here, where Midy does not work, the number of nines is a quarter of the period length.

(2) Beginning with period length 24, every 6-digit sum is repeated

　　a) Twice (24), 4 times (48), and 8 times (96)

　　b) Those numbers (2, 4, and 8) are all powers of two.

(3) Every period length is a multiple of eight.

(4) Every 6-digit sum adds up to 27, because this makes possible the Midy effect in the last column.

(5) Two numbers, 511 and 119, have the same repetend total. They both have 7 as one factor, and 73 or 17 as the other factor; 73 and 17 are both numbers one greater than a multiple of eight.

Table 2

Adding More Than Two Digits

DIVISOR OF ONE	PERIOD LENGTH	REPETEND HALVES – 1*	REPETEND HALVES - 2**
1241 = 17 × 73	16	00080580 17727639 17808219	1780 8219 9999
511 = 7 × 73	24	001956 947162 426614 481409 428571,428571	428 571 999
949 = 13 × 73	24	001053 740779 768177 028451 769230,769230	769 230 999
119 = 7 × 17	48	008403 361344 537815 126050 420168 067226 890756 302521 428571,428571,428571,428571	428 571 999
221 = 13 × 17	48	004524 886877 828054 298642 533936 651583 710407 239819 538461,538461,538461,538461	538 461 999
679 = 7 × 97	96	001472 754050 073637 702503 681885 125184 094256 259204 712812 960235 640648 011782 032400 589101 620029 455081 714285,714285,714285,714285,714285,714285,714285,714285	714 285 999
1261 = 13 × 97	96	000793 021411 578112 609040 444091 990483 743061 062648 691514 670896 114195 083267 248215 701823 949246 629659 692307,692307,692307,692307,692307,692307,692307,692307	692 307 999

* Halves of repetend added.   ** Halves of the sum of repetend halves added.

19

References

1. Balmond C. *Number 9: The Search for the Sigma Code.* Prestel/Verlag: Munich, Germany; 1998; 232 pp.

2. Midy E. De quelques propriétés des nombres et. des fractions décimales périodiques. *Nantes, France*; 1836; 21 pp.

3. Yates S. *Repunits and Repetends.* Star Publishing Co: Boynton Beach, FL; 1982; p 27.

4. Lewittes J. Midy's theorem for periodic decimals. <u>In</u> *Integers: Electronic Journal of Combinatorial Number Theory 7*; published online January 23, 2007 (#AO2); 11 pp. At: www.emis.de/journals/INTEGERS/papers/h2/h2.pdf. Accessed 11/29/13.

MAGIC SQUARES

by Clarence A. Gipbsin

Magic sum 333

114	107	112
109	111	113
110	115	108

Magic sum 555

115	122	99	106	113
121	103	105	112	114
102	104	111	118	120
108	110	117	119	101
109	116	123	100	107

Magic sum 777

116	125	134	87	96	105	114
124	133	93	95	104	113	115
132	92	94	103	112	121	123
91	100	102	111	120	122	131
99	101	110	119	128	130	90
107	109	118	127	129	89	98
108	117	126	135	88	97	106

Magic sum 666

80	91	102	113	34	45	56	67	78
90	101	112	42	44	55	66	77	79
100	111	41	43	54	65	76	87	89
110	40	51	53	64	75	86	88	99
39	50	52	63	74	85	96	98	109
49	60	62	73	84	95	97	108	38
59	61	72	83	94	105	107	37	48
69	71	82	93	104	106	36	47	58
70	81	92	103	114	35	46	57	68

Magic sum 999

117	128	139	150	71	82	93	104	115
127	138	149	79	81	92	103	114	116
137	148	78	80	91	102	113	124	126
147	77	88	90	101	112	123	125	136
76	87	89	100	111	122	133	135	146
86	97	99	110	121	132	134	145	75
96	98	109	120	131	142	144	74	85
106	108	119	130	141	143	73	84	95
107	118	129	140	151	72	83	94	105

SMARANDACHE'S ORTHIC THEOREM

Edited by Prof. Ion Patrascu

Fratii Buzesti College

Craiova, Romania

Abstract

We present the Smarandache's Orthic Theorem in the geometry of the triangle.

Smarandache's Orthic Theorem

Given a triangle ABC whose angles are all acute (acute triangle), we consider $A'B'C'$, the triangle formed by the legs of its altiudes.

What are the conditions where the expression

$$\|A'B'\| \cdot \|B'C'\| + \|B'C'\| \cdot \|C'A'\| + \|C'A'\| \cdot \|A'B'\|$$

is maximum?

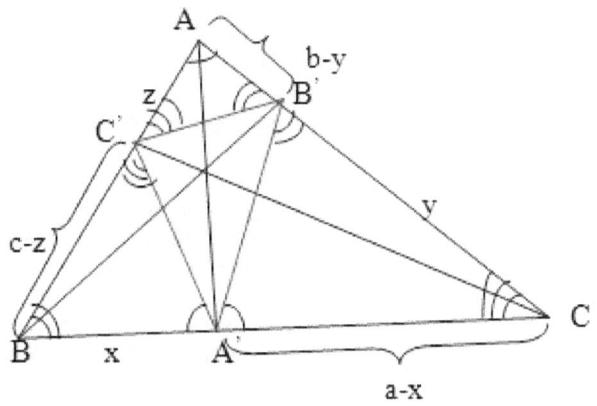

23

Solution:

We have

$$\Delta ABC \sim \Delta A'B'C' \sim \Delta AB'C \sim \Delta A'BC' \tag{1}$$

We note

$$\|BA'\| = x, \quad \|CB'\| = y, \quad \|AC'\| = z .$$

It follows that

$$\|A'C\| = a - x, \quad \|B'A\| = b - y, \quad \|C'B\| = c - z$$

$$B\widehat{A}C = B'\widehat{A'}C = B\widehat{A'}C'; \quad A\widehat{B}C = A\widehat{B'}C' = A'\widehat{B'}C; \quad B\widehat{C}A = B\widehat{C'}A' = B'\widehat{C'}A$$

From these equalities we have the relation (1)

$$\Delta A'BC' \sim \Delta A'B'C \Rightarrow \frac{\|A'C'\|}{a - x} = \frac{x}{\|A'B'\|} \tag{2}$$

$$\Delta A'B'C \sim \Delta AB'C' \Rightarrow \frac{\|A'C'\|}{z} = \frac{c - z}{\|B'C'\|} \tag{3}$$

$$\Delta AB'C' \sim \Delta A'B'C \Rightarrow \frac{\|B'C'\|}{y} = \frac{b - y}{\|A'B'\|} \tag{4}$$

From (2), (3) and (4) we observe that the sum of the products from the problem is equal to:

$$x(a - x) + y(b - y) + z(c - z) = \frac{1}{4}\left(a^2 + b^2 + c^2\right) - \left(x - \frac{a}{2}\right)^2 - \left(y - \frac{b}{2}\right)^2 - \left(z - \frac{c}{2}\right)^2$$

24

which will reach its maximum when $x = \dfrac{a}{2}$, $y = \dfrac{b}{2}$, $z = \dfrac{c}{2}$, that is when the altitudes' legs are in

the middle of the sides, therefore when $\triangle ABC$ is equilateral. The maximum of the expression is

$$\frac{1}{4}\left(a^2 + b^2 + c^2\right).$$

Conclusion (Smarandache's Orthic Theorem)

If we note the lengths of the sides of the triangle $\triangle ABC$ by $\|AB\| = c$, $\|BC\| = a$, $\|CA\| = b$, and

the lengths of the sides of its orthic triangle $\triangle A`B`C`$ by $\|A`B`\| = c`$, $\|B`C`\| = a`$, $\|C`A`\| = b`$,

then we have proved that:

$$4(a`b` + b`c` + c`a`) \le a^2 + b^2 + c^2.$$

Open Problems related to Smarandache's Orthic Theorem:

1. Generalize this problem to polygons. Let $A_1A_2\ldots A_m$ be a polygon and P a point inside it.
   From P we draw perpendiculars on each side A_iA_{i+1} of the polygon and we note by Ai'
   the intersection between the perpendicular and the side A_iA_{i+1}. A pedal polygon
   $A_1`A_2`\ldots A_m`$ is formed. What properties does this pedal polygon have?

2. Generalize this problem to polyhedrons. Let $A_1A_2\ldots A_n$ be a poliyhedron and P a point
   inside it. From P we draw perpendiculars on each polyhedron face F_i and we note by Ai'
   the intersection between the perpendicular and the side F_i. A pedal polyhedron
   $A_1`A_2`\ldots A_p`$ is formed, where p is the number of polyhedron's faces. What properties
   does this pedal polyhedron have?

References

1. Cătălin Barbu, *Teorema lui Smarandache*, in his book "Teoreme fundamentale din geometria triunghiului", Chapter II (Teoreme fundamentale din geometria triunghiului), Section II.57, p. 337, Editura Unique, Bacău, 2008.

2. József Sándor, *On Smarandache's Pedal Theorem*, in his book *Geometric Theorems, Diophantine Equations, and Arithmetic Functions*, AR Press, pp. 9-10, Rehoboth, 2002.

3. Ion Pătrașcu, *Smarandache's Orthic Theorem*,
http://www.scribd.com/doc/28311593/Smarandache-s-Orthic-Theorem

4. F. Smarandache, *Eight Solved and Eight Open Problems in Elementary Geometry*, in arXiv.org, Cornell University, NY, USA.

5. F. Smarandache, *Problèmes avec et sans… problèmes!*, Problem 5.41, p. 59, Somipress, Fés, Morocco, 1983.

Exploring a Fascinating Integer Sequence

Jay L. Schiffman

Mathematics Department

Rowan University

Glassboro, NJ 08028 USA

schiffman@rowan.edu

Abstract

A number of years ago, a colleague asked his students to find all the prime outputs in the following integer sequence: 9, 98, 987, 9876, 98765, 987654, 9876543, 98765432, 987654321, 9876543219, 98765432198, 987654321987, etc. This article not only resolves this question, but pursues the sequence much further. Using MATHEMATICA and modular arithmetic, we will explore the prime factorizations and divisibility patterns in the sequence as well as suggest companion sequences and directions for additional stimulating research.

Introduction

Our initial goal is to show that there are no primes in this sequence. Observe that the first term 9 is divisible by 3, the terms 98, 9876, 987654 and 98765432 are even and the term 98765 is divisible by 5. It is well known that any integer is divisible by 3 or 9 if and only if the integer obtained by forming the digital sum is divisible by 3 or 9. As a consequence, the terms 987, 9876543, 987654321 and 9876543219 are divisible by both 3 and 9. If one appends the digits 987, 9876, 987654, 9876543 and 987654321 to the integer 987654321 then divisibility by 3 is preserved. Clearly 98765432198765 is divisible by 5 while both 98765432198 and 98765432198765432 are divisible by 2. Hence there are no primes in this sequence. In order to view divisibility patterns, it is helpful to factor the integers in question.

Factorization of the Integers

Our first goal is to generate the complete factorizations for the initial one hundred nine terms which represents the number of iterations I have achieved using MATHEMATICA Version 9.0. The purpose of these factorizations is to enable one to view patterns and form conjectures based on the analysis of these patterns and then attempt to affirm the conjectures with formal proofs.

Table 1

The complete factorizations of the initial one hundred nine iterations are now furnished

$1st : 9 :$

3^2

$2nd : 98 :$

$2 \cdot 7^2$

$3rd : 987 :$

$3 \cdot 7 \cdot 47$

$4th$: 9876 :

$2^2 \cdot 3 \cdot 823$

$5th$: 98765 :

$5 \cdot 19753$

$6th$: 987654 :

$2 \cdot 3 \cdot 97 \cdot 1697$

$7th$: 9876543 :

$3 \cdot 227 \cdot 14503$

$8th$: 98765432 :

$2^3 \cdot 37 \cdot 333667$

$9th$: 987654321 :

$3^2 \cdot 17^2 \cdot 379721$

$10th$: 9876543219 :

$3^3 \cdot 15391 \cdot 23767$

$11th$: 98765432198 :

$2 \cdot 19 \cdot 23 \cdot 73 \cdot 1061 \cdot 1459$

$12th$: 987654321987 :

$3 \cdot 329281 \cdot 999809$

$13th$: 9876543219876 :

$2^2 \cdot 3 \cdot 19 \cdot 40129 \cdot 1079473$

$14th$: 98765432198765 :

$5 \cdot 11329 \cdot 1743586057$

$15th$: 987654321987654 :

$2 \cdot 3 \cdot 61 \cdot 257 \cdot 11071 \cdot 948427$

$16th$: 9876543219876543 :

$3 \cdot 7 \cdot 79 \cdot 5953311163277$

$17th$: 98765432198765432 :

$2^3 \cdot 127 \cdot 97210071061777$

18*th* : 987654321987654321 :

$3^2 \cdot 7 \cdot 11 \cdot 13 \cdot 17^2 \cdot 19 \cdot 52579 \cdot 379721$

19*th* : 9876543219876543219 :

$3^2 \cdot 67 \cdot 83 \cdot 311 \cdot 634525638821$

20*th* : 98765432198765432198 :

$2 \cdot 7 \cdot 350887 \cdot 20105258184211$

21*st* : 987654321987654321987 :

$3 \cdot 7 \cdot 185470889 \cdot 253577035423$

22*nd* : 9876543219876543219876 :

$2^2 \cdot 3 \cdot 607 \cdot 297461987 \cdot 4558306847$

23*rd* : 98765432198765432198765 :

$5 \cdot 1123 \cdot 17589569403163923811$

24*th* : 987654321987654321987654 :

$2 \cdot 3 \cdot 301579 \cdot 203737187 \cdot 2679059233$

25*th* : 9876543219876543219876543 :

$3 \cdot 53 \cdot 11540561 \cdot 22319939 \cdot 241150363$

26*th* : 98765432198765432198765432 :

$2^3 \cdot 645750629 \cdot 19118338365328451$

27*th* : 987654321987654321987654321 :

$3^3 \cdot 17^2 \cdot 757 \cdot 379721 \cdot 440334654777631$

28*th* : 9876543219876543219876543219 :

$3^2 \cdot 967 \cdot 1134843527505060693999373$

29*th* : 98765432198765432198765432198 :

$2 \cdot 19 \cdot 179 \cdot 191 \cdot 3323 \cdot 21347 \cdot 1071686555248069$

30*th* : 987654321987654321987654321987 :

$3 \cdot 79 \cdot 193 \cdot 810361 \cdot 19516817 \cdot 1365248763511$

31*st* : 9876543219876543219876543219876 :

$2^2 \cdot 3 \cdot 19^2 \cdot 321371 \cdot 69864647 \cdot 101543555831239$

$32nd$: $9876543219876543219876543219876 5$:
$5 \cdot 31 \cdot 622499921 \cdot 5390335409 \cdot 189897033367$

$33rd$: $98765432198765432198765432198765 4$:
$2 \cdot 3 \cdot 1702219 \cdot 9670262972309030369453581 1$

$34th$: $987654321987654321987654321987654 3$:
$3 \cdot 7 \cdot 458639 \cdot 5792219 \cdot 91191293 \cdot 194140589709 1$

$35th$: $9876543219876543219876543219876543 2$:
$2^{3} \cdot 61 \cdot 20238818073517506598107670532533 9$

$36th$: $98765432198765432198765432198765432 1$:
$3^{2} \cdot 7 \cdot 11 \cdot 13 \cdot 17^{2} \cdot 19 \cdot 101 \cdot 9901 \cdot 52579 \cdot 379721 \cdot 99999900000 1$

$37th$: $9876543219876543219876543219876543 219$:
$3^{6} \cdot 53 \cdot 827 \cdot 2663925701 \cdot 3048017789 \cdot 38067684029$

$38th$: $98765432198765432198765432198765432 198$:
$2 \cdot 7 \cdot 229 \cdot 4402693363 \cdot 6997179443486948677576891$

$39th$: $987654321987654321987654321987654 3219 87$:
$3 \cdot 7 \cdot 131 \cdot 3089 \cdot 116224174403357644453802233954733$

$40th$: $9876543219876543219876543219876543 2198 76$:
$2^{2} \cdot 3 \cdot 37 \cdot 283 \cdot 3889 \cdot 6449 \cdot 3134045107237960056786308933$

$41st$: $98765432198765432198765432198765432 19876 5$:
$5 \cdot 179 \cdot 349 \cdot 5557 \cdot 480167 \cdot 11850148701114908782078539 7$

$42nd$: $987654321987654321987654321987654321 987654$:
$2 \cdot 3 \cdot 769 \cdot 105337 \cdot 20321063497152924204344072786565 53$

$43rd$: $9876543219876543219876543219876543 2198 76543$:
$3 \cdot 161641 \cdot 20367240200767015010375963235145668941$

$44th$: $98765432198765432198765432198765432 19876 5432$:
$2^{3} \cdot 2245766936983901 \cdot 5497310883660132998369867579$

31

$45th$: $9876543219876543219876543219876543219876654321$:
$3^2 \cdot 17^2 \cdot 31 \cdot 41 \cdot 271 \cdot 238681 \cdot 379721 \cdot 2906161 \cdot 4185502830133110721$

$46th$: $9876543219876543219876543219876543219876543219$:
$3^2 \cdot 23 \cdot 47 \cdot 3037 \cdot 47797 \cdot 54368184863 \cdot 132527925857 \cdot 970597346189$

$47th$: $9876543219876543219876543219876543219876543219\,8$:
$2 \cdot 19 \cdot 886836055497204325459 \cdot 2930744983708362950897419$

$48th$: $9876543219876543219876543219876543219876543219\,87$:
$3 \cdot 2039 \cdot 2281 \cdot 70784994520316800756407035050022609816431$

$49th$: $9876543219876543219876543219876543219876543219\,876$:
$2^2 \cdot 3 \cdot 19 \cdot 97 \cdot 809 \cdot 248891 \cdot 1033457 \cdot 92929628123 \cdot 23093728862268520129$

$50th$: $9876543219876543219876543219876543219876543219\,8765$:
$5 \cdot 701 \cdot 6606343 \cdot 78513403 \cdot 54326537537257274953710732487457$

$51st$: $9876543219876543219876543219876543219876543219\,87654$:
$2 \cdot 3 \cdot 164609053664609053664609053664609053664609053664609$

$52nd$: $9876543219876543219876543219876543219876543219\,876543$:
$3 \cdot 7 \cdot 1122151805535999089 \cdot 419115826912774345231822578935437$

$53rd$: $9876543219876543219876543219876543219876543219\,8765432$:
$2^3 \cdot 97 \cdot 743 \cdot 3911767 \cdot 6901767894361 \cdot 6344846284516354997149949927$

54*th* : 98765432198765432198765432198765432198765432198765 4321 :
$3^3 \cdot 7 \cdot 11 \cdot 13 \cdot 17^2 \cdot 19 \cdot 757 \cdot 52579 \cdot 379721 \cdot 70541929 \cdot 14175966169 \cdot 440334654777631$

55*th* : 98765432198765432198765432198765432198765432198765 43219 :
$3^2 \cdot 24353244346939243 \cdot 45061498807460369813224481945700561937$

56*th* : 98765432198765432198765432198765432198765432198765 432198 :
$2 \cdot 7 \cdot 7054673728483245157054673728483245157054673728483245157$

57*th* : 98765432198765432198765432198765432198765432198765 4321987 :
$3 \cdot 7 \cdot 470311581898883010470311581898883010470311581898883 01047$

58*th* : 98765432198765432198765432198765432198765432198765 43219876 :
$2^2 \cdot 3 \cdot 478912843 \cdot 17185700495466672384124083060412837337983 96524361$

59*th* : 98765432198765432198765432198765432198765432198765 432198765 :
$5 \cdot 1871 \cdot 466752507663839363 \cdot 22619058135403591340614270026541184461$

60*th* : 98765432198765432198765432198765432198765432198765 4321987654 :
$2 \cdot 3 \cdot 251 \cdot 655812962807207385117964357229518142090075911014378699859$

61*st* : 98765432198765432198765432198765432198765432198765 43219876543 :
$3 \cdot 23 \cdot 359 \cdot 89733521 \cdot 44433110850950239431249355675146428662464 44183173$

62*nd* : 98765432198765432198765432198765432198765432198765 4321987654 32 :
$2^3 \cdot 59 \cdot 14011 \cdot 14934608309990913948780775183718457319667330420584406471$

$63rd$: 9876543219876543219876543219876543219876543219876543219876543219 :

$3^2 \cdot 17^2 \cdot 43 \cdot 239 \cdot 1933 \cdot 4649 \cdot 10837 \cdot 23311 \cdot 45613 \cdot 379721 \cdot 10838689 \cdot 45121231 \cdot 1921436048294281$

$64th$: 98765432198765432198765432198765432198765432198765432198765432198765432 19 :

$3^3 \cdot 397 \cdot 3109 \cdot 174173666719047629352227771 \cdot 170156083514676818862243186685 9$

$65th$: 98765432198765432198765432198765432198765432198765432198765432198765432198 :

$2 \cdot 19 \cdot 314441 \cdot 182024794087 \cdot 4117556065279609753 \cdot 11028390448163680720577786471$

$66th$: 98765432198765432198765432198765432198765432198765432198765432198765432198 7 :

$3 \cdot 61 \cdot 89 \cdot 19158431 \cdot 377475370741543277323 \cdot 838523643335828951272007021411197 7$

$67th$: 98765432198765432198765432198765432198765432198765432198765432198765432198 76 :

$2^2 \cdot 3 \cdot 19 \cdot 8682356617468691775883 \cdot 4989218241721063020618157618293818811101299$

$68th$: 98765432198765432198765432198765432198765432198765432198765432198765432198 765 :

$5 \cdot 53 \cdot 385691049284423 \cdot 10711784976913073843 \cdot 902106242284171325365600831094 09$

$69th$: 98765432198765432198765432198765432198765432198765432198765432198765432198 7654 :

$2 \cdot 3 \cdot 32069 \cdot 59549117 \cdot 8619716344911828025687655345555209663287911168254952843 3$

$70th$: 98765432198765432198765432198765432198765432198765432198765432198765432198 76543 :

$3 \cdot 7^2 \cdot 3757673 \cdot 11055130505084839 \cdot 1617352841157561666891489825930280730671806427$

$71st$: 98765432198765432198765432198765432198765432198765432198765432198765432198 765432 :

$2^3 \cdot 17981 \cdot 484601729 \cdot 1173956074878835543 \cdot 12068806761944499832756243896831727191 97$

$72nd$: 9876543219876543219876543219876543219876543219876543219876543219876543219876 54321 :

$3^2 \cdot 7 \cdot 11 \cdot 13 \cdot 17^2 \cdot 19 \cdot 73 \cdot 101 \cdot 137 \cdot 3169 \cdot 9901 \cdot 52579 \cdot 98641 \cdot 379721 \cdot 99990001 \cdot 999999000001 \cdot 3199044596370769$

73rd : 9876543219876543219876543219876543219876543219876543219876543219876543219 :

$3^2 \cdot 6684983042950796671 \cdot 1880178439766481446489 \cdot 873098175882218256939326196269 89$

74th : 9876543219876543219876543219876543219876543219876543219876543219876543219 8 :

$2 \cdot 7 \cdot 4149561914700601669865687950883 \cdot 17001008476318282653930641723030429 38193879$

75th : 9876543219876543219876543219876543219876543219876543219876543219876543219 87 :

$3 \cdot 7 \cdot 6053 \cdot 281663 \cdot 2140903 \cdot 3134246417 \cdot 41110718818338714760515475180758845835270 90201523$

76th : 9876543219876543219876543219876543219876543219876543219876543219876543219 876 :

$2^2 \cdot 3 \cdot 23 \cdot 29 \cdot 57901 \cdot 2074243 \cdot 18875231 \cdot 5443270485229901765722456792233065507886095660 00406593$

77th : 9876543219876543219876543219876543219876543219876543219876543219876543219 8765 :

$5 \cdot 31 \cdot 79 \cdot 97 \cdot 6115068017 \cdot 8670352473784685605243410415 19 \cdot 156832620726946316829552706516 87$

78th : 9876543219876543219876543219876543219876543219876543219876543219876543219 87654 :

$2 \cdot 3 \cdot 619 \cdot 220379869 \cdot 761355641949193 \cdot 66660001482131287 \cdot 23775970827715189218208511178777 809$

79th : 9876543219876543219876543219876543219876543219876543219876543219876543219 876543 :

$3 \cdot 71 \cdot 157 \cdot 295342340835397961181679471902052666483554415833155205282 3231458983778691423$

80th : 9876543219876543219876543219876543219876543219876543219876543219876543219 8765432 :

$2^3 \cdot 3866143 \cdot 3327658297 \cdot 52139070875496169377457 \cdot 18404967555399486945368824092669657 134857$

81st : 9876543219876543219876543219876543219876543219876543219876543219876543219 87654321 :

$3^4 \cdot 17^2 \cdot 163 \cdot 757 \cdot 9397 \cdot 379721 \cdot 2462401 \cdot 440334654777631 \cdot 676421558270641 \cdot 1306548978080077784250 46117$

82nd : 9876543219876543219876543219876543219876543219876543219876543219876543219 876543219 :

$3^2 \cdot 435437 \cdot 1500400199 \cdot 2373821122669 \cdot 21007502066399868934292 9 \cdot 33682755434731187797063458311 57$

83*rd* : 98765432198765432198765432198765432198765432198765432198765432198765432198 :
2·19·73·107·157·434561·1504904963·324082260644981847006636258249476647840782901130870521387126 1

84*th* : 98765432198765432198765432198765432198765432198765432198765432198765432198765 43219 87 :
3·6388619·344915047·1494048115620554206917314509547025220524565372567184720116368881468 53

85*th* : 98765432198765432198765432198765432198765432198765432198765432198765432198765432198 76 :
2^2·3·19·300683·795825959·1810269130126407410904563416125809518398142911986250947728961621526 61

86*th* : 987654321987654321987654321987654321987654321987654321987654321987654321987654321987 65 :
5·23·433·1268563682741·11890171712999·1314979042488475027511098623524731352184902821206649138 13

87*th* : 987654321987654321987654321987654321987654321987654321987654321987654321987654321987 654 :
2·3·746117·2148617·112771378590551·9105187336221266255132758688269235038837006103793541478817 31

88*th* : 987654321987654321987654321987654321987654321987654321987654321987654321987654321987 6543 :
3·7·28661·285266064908292323070264776350 3·5752335704316649038503470462495265711025899605291801

89*th* : 987654321987654321987654321987654321987654321987654321987654321987654321987654321987 65432 :
2^3·29·359287847·419823631·28223285945541840245821274785125547377099754561274723720695602892206 43

90*th* : 987654321987654321987654321987654321987654321987654321987654321987654321987654321987 654321 :
3^2·7·11·13·17^2·19·31·41·211·241·271·2161·9091·29611·52579·238681·379721·2906161·3762091·8985695684401·4185502830133110721

91*st* : 987654321987654321987654321987654321987654321987654321987654321987654321987654321987 6543219 :
3^3·29983·103527718180699·6153313410572947·6825007988650399·2806062098299295061674240243009291644297

92*nd* : 987654321987654321987654321987654321987654321987654321987654321987654321987654321987 65432198 :
2·7·10939·2500709117·47569402356620429·54213623820285520259833598046440435457442199290571105 49193591

93*rd* : 987654321987654321987654321987654321987654321987654321987654321987654321987654321987 654321987 :
3·7·53·137992199·856736557·57971981387·1294761645748109083222403609870659454172602214914650833 03202339

36

94*th* : 9876543219876543219876543219876543219876543219876543219876543219876543219876543219876 :
$2^2 \cdot 3 \cdot 419 \cdot 2029346660387524751161995882865 3 \cdot 9679511907093352602358905594187316830264070953649075178 9989$

95*th* : 9876543219876543219876543219876543219876543219876543219876543219876543219876543219876 5 :
$5 \cdot 47 \cdot 313 \cdot 2833 \cdot 16763 \cdot 2068427 \cdot 5486084413 \cdot 308798378179149932501597 9 \cdot 806894498872415880009950767600368610205353$

96*th* : 9876543219876543219876543219876543219876543219876543219876543219876543219876543219876 54 :
$2 \cdot 3 \cdot 283 \cdot 97231 \cdot 7012225249 \cdot 9046429879 \cdot 960014530549 \cdot 6692559825509 \cdot 66031069549997 \cdot 222285457217721330672724298399$

97*th* : 9876543219876543219876543219876543219876543219876543219876543219876543219876543219876 543 :
$3 \cdot 61 \cdot 183493487 \cdot 5216280375646633678304681383195633055396 57 \cdot 563861312968950465023116091206616965834636319$

98*th* : 9876543219876543219876543219876543219876543219876543219876543219876543219876543219876 5432 :
$2^3 \cdot 1741 \cdot 1236239 \cdot 7076803 \cdot 16958573 \cdot 17963119319254379 \cdot 3742148303841258867894808 9 \cdot 711024571419228606634112544066 89$

99*th* : 9876543219876543219876543219876543219876543219876543219876543219876543219876543219876 54321 :
$3^2 \cdot 17^2 \cdot 67 \cdot 199 \cdot 397 \cdot 21649 \cdot 34849 \cdot 379721 \cdot 513239 \cdot 1344628210313298373 \cdot 36285372434299046932476623547426886978631188605388 3$

100*th* : 9876543219876543219876543219876543219876543219876543219876543219876543219876543219876 54321 9 :
$3^2 \cdot 1097393691097393691097393691097393691097393691097393691097393691097393691097393691097393691097393691$

101*st* : 98765432198765432198765432198765432198765432198765432198765432198765432198765432198765432198 :
$2 \cdot 19 \cdot 83 \cdot 16183 \cdot 108517 \cdot 39920213 \cdot 111577871 \cdot 15496138811 \cdot 25834023368262517413544714768829686641566387657092011734540068 9$

102*nd* : 98765432198765432198765432198765432198765432198765432198765432198765432198765432198765432198 7 :
$3 \cdot 409 \cdot 15238086408007 \cdot 52823840618579183084913479817960034755411349570659126331041848003680701890744952459983$

103*rd* : 98765432198765432198765432198765432198765432198765432198765432198765432198765432198765432198765432 1987 6 :
$2^2 \cdot 3 \cdot 19 \cdot 3485440640491 \cdot 19350464058639199 \cdot 6422750418773635958405001106203098352870710949512909672594525415402721 3$

104*th* : 98765432198765432198765432198765432198765432198765432198765432198765432198765432198765432198765432 19876 5 :
$5 \cdot 22328946883 \cdot 15370065351726819739 \cdot 57556053757186951652000935065199886989005188488670205743194438455012261 69$

37

105th: 9876543219876543 :

$2 \cdot 3 \cdot 487 \cdot 166076101 \cdot 2035249312353137139593799296959797606121155404115695570389646116214742279617369041613391664907$

106th: 9876543219876543219876543219876543219876543219876543219876543219876543219876543219876543219876543219876543219876543219876543219876543219876543219876543 :

$3 \cdot 7 \cdot 16925714308848825495186373283 \cdot 27786808480691558828717035722586942088356228059911248255076402571976905943201$

107th : 98765432198765432198765432198765432198765432198765432198765432198765432198765432198765432198765432198765432198765432198765432198765432198765432198765432198765432 :

$2^{3} \cdot 691 \cdot 11411 \cdot 1138991209 \cdot 288711920815409823773 \cdot 4761328441375758245182359455236779663252620639153503524090320670837547$

108th: 987654321 :

$3^{3} \cdot 7 \cdot 11 \cdot 13 \cdot 17^{2} \cdot 19 \cdot 101 \cdot 109 \cdot 757 \cdot 9901 \cdot 52579 \cdot 153469 \cdot 379721 \cdot 70541929 \cdot 14175966169 \cdot 999999000001 \cdot 440334654777631 \cdot 59779577156334533866654838281$

109th : 9876543219876543219876543219876543219876543219876543219876543219876543219876543219876543219876543219876543219876543219876543219876543219876543219876543219876543219876543219 :

$3^{2} \cdot 94351 \cdot 8690069 \cdot 3272974869029576819908556441 \cdot 408931020348682352325098350872802337769558386546700954373032787501129$

Some Divisibility Conjectures

In order to simplify our notation, let us agree to call $S(1)$, $S(2)$,..., $S(k)$ the first, second and

$k-th$ terms of the sequence respectively. Of course, $S(1)=9$, $S(2)=98$, $S(3)=987$, etc.

Based on the analysis of the factorizations, the following conjectures can be established and

justified where $k \in W = \{0,1,2,3,4,5,...\}$, the set of Whole Numbers:

Conjecture 1: $S(k)$ is even if and only if $k \equiv 2, 4, 6 \text{ or } 8 \pmod{9}$.

Conjecture 2: $S(k)$ is divisible by 3 if and only if $k \equiv 0, 1, 3, 4, 6 \text{ or } 7 \pmod{9}$.

More simply, $S(k)$ is divisible by 3 if and only if $k \not\equiv 2 \pmod{3}$.

Conjecture 3: $S(k)$ is divisible by 4 if and only if $k \equiv 4 \text{ or } 8 \pmod{9}$.

38

Conjecture 4: $S(k)$ is divisible by 5 if and only if $k \equiv 5 \pmod{9}$.

Conjecture 5: $S(k)$ is divisible by 6 if and only if $k \equiv 4$ *or* $6 \pmod{9}$.

Conjecture 6: $S(k)$ is divisible by 7 if and only if $k \equiv 0, 2, 3$ *or* $16 \pmod{18}$.

Conjecture 7: $S(k)$ is divisible by 8 if and only if $k \equiv 8 \pmod{9}$.

Conjecture 8: $S(k)$ is divisible by 9 if and only if $k \equiv 0$ *or* $1 \pmod{9}$.

Conjecture 9: $S(k)$ is never divisible by any one of the following integers: 10, 15, 16, 18, 20, 22, 24, 25, 26, 28, 32, 34, 35, 36, 40, 42, 44, 45, 48, 50, 54, 55, 56, 60, 62, 64, 65, 66, 68, 70, 72, 75, 78, 80, 84, 85, 86, 87, 88, 96 or 100.

Conjecture 10: $S(k)$ is divisible by 11 if and only if $k \equiv 0 \pmod{18}$.

Conjecture 11: $S(k)$ is divisible by 12 if and only if $k \equiv 4 \pmod{9}$.

Conjecture 12: $S(k)$ is divisible by 13, 33, 39, 57, 63, 77, 91 and 99 if and only if $k \equiv 0 \pmod{18}$.

Conjecture 13: $S(k)$ is divisible by 14 if and only if $k \equiv 2 \pmod{18}$.

Conjecture 14: $S(k)$ is divisible by 19 if and only if $k \equiv 0, 11$ *or* $13 \pmod{18}$.

Conjecture 15: $S(k)$ is divisible by 21 if and only if $k \equiv 0, 3$ *or* $16 \pmod{18}$.

Conjecture 16: $S(k)$ is divisible by 27 if and only if $k \equiv 0$ *or* $10 \pmod{27}$.

Conjecture 17: $S(k)$ is divisible by 17 and 51 if and only if $k \equiv 0 \pmod 9$.

In order to prove these true conjectures, one appeals to modular arithmetic. For example, let us consider a test for divisibility of an integer by 13. One notes the following:

$$1 \equiv 1 \pmod{13}$$
$$10 \equiv 10 \pmod{13} \equiv -3 \pmod{13}$$
$$10^2 \equiv 9 \pmod{13} \equiv -4 \pmod{13}$$
$$10^3 \equiv 12 \pmod{13} \equiv -1 \pmod{13}$$
$$10^4 \equiv 3 \pmod{13}$$
$$10^5 \equiv 4 \pmod{13}$$
$$10^6 \equiv 1 \pmod{13}$$

$$N = a_0 \cdot 10^0 + a_1 \cdot 10^1 + a_2 \cdot 10^2 + a_3 \cdot 10^3 + a_4 \cdot 10^4 + a_5 \cdot 10^5 + a_6 \cdot 10^6 + \ldots \equiv a_0 - 3 \cdot a_1 - 4 \cdot a_2 - 1 \cdot a_3 + 3 \cdot a_4 + 4 \cdot a_5 + 1 \cdot a_6 + \ldots \pmod{13}.$$

In essence, we are asserting that to test whether an integer is divisible by 13, proceeding from right to left, examine the integer obtained by taking the units digit minus three times the tens digit minus four times the hundreds digit minus the thousands digit plus three times the ten thousands digit plus four times the hundreds thousands digit plus the millions digit, etc. If the result is congruent to 0 mod 13, then the integer is divisible by 13. The sequence of multipliers $(1, -3, -4, -1, 3, 4, \ldots)$ repeats in that pattern. It is useful to view this on a six hour analogue clock which the reader is invited to partake of.

To cite an example, $S(18) = 987654321987654321$ is divisible by 13; for

$$1 \cdot 1 - 3 \cdot 2 - 4 \cdot 3 - 1 \cdot 4 + 3 \cdot 5 + 4 \cdot 6 + 1 \cdot 7 - 3 \cdot 8 - 4 \cdot 9 - 1 \cdot 1 + 3 \cdot 2 + 4 \cdot 3 + 1 \cdot 4 - 3 \cdot 5 - 4 \cdot 6 - 1 \cdot 7 + 3 \cdot 8 + 4 \cdot 9 =$$
$$1 - 6 - 12 - 4 + 15 + 24 + 7 - 24 - 36 - 1 + 6 + 12 + 4 - 15 - 24 - 7 + 24 + 36 = 0 \equiv 0 \pmod{13}.$$

We next test an integer for divisibility by 27. Observe the following:

40

$$1 \equiv 1 \pmod{27}$$
$$10 \equiv 10 \pmod{27}$$
$$10^2 \equiv 19 \pmod{27} \equiv -8 \pmod{27}$$
$$10^3 \equiv 1 \pmod{27}$$

$$N = a_0 \cdot 10^0 + a_1 \cdot 10^1 + a_2 \cdot 10^2 + a_3 \cdot 10^3 + \ldots \equiv a_0 + 10 \cdot a_1 - 8 \cdot a_2 + 1 \cdot a_3 + \ldots \pmod{27}.$$

To cite an example, observe that $S(10) = 9876543219$ and

$S(27) = 987654321987654321987654321$ is each divisible by 27.

For $S(10)$, note that

$$1 \cdot 9 + 10 \cdot 1 - 8 \cdot 2 + 1 \cdot 3 + 10 \cdot 4 - 8 \cdot 5 + 1 \cdot 6 + 10 \cdot 7 - 8 \cdot 8 + 1 \cdot 9 =$$
$$9 + 10 - 16 + 3 + 40 - 40 + 6 + 70 - 64 + 9 = 27 \equiv 0 \pmod{27}.$$

For $S(27)$, observe

$$1 \cdot 1 + 10 \cdot 2 - 8 \cdot 3 + 1 \cdot 4 + 10 \cdot 5 - 8 \cdot 6 + 1 \cdot 7 + 10 \cdot 8 - 8 \cdot 9 + 1 \cdot 1 + 10 \cdot 2 - 8 \cdot 3 + 1 \cdot 4 + 10 \cdot 5 - 8 \cdot 6 + 1 \cdot 7 +$$
$$10 \cdot 8 - 8 \cdot 9 + 1 \cdot 1 + 10 \cdot 2 - 8 \cdot 3 + 1 \cdot 4 + 10 \cdot 5 - 8 \cdot 6 + 1 \cdot 7 + 10 \cdot 8 - 8 \cdot 9 = 1 + 20 - 24 + 4 + 50 - 48 + 7 +$$
$$80 - 72 + 1 + 20 - 24 + 4 + 50 - 48 + 7 + 80 - 72 + 1 + 20 - 24 + 4 + 50 - 48 + 7 + 80 - 72 = 54 \equiv 0 \pmod{27}.$$

These two computations aid in establishing the truth of **Conjecture 16.**

If one examines the structure of the sequence, the highest power of two that is possible as a

factor of any term is 2^3. Note that an integer is divisible by two if and only if the units digit is

even which accounts for the second, fourth, sixth, eighth, eleventh, thirteenth, fifteenth,

seventeenth, etc. terms being even. If we code the terms of the sequence by Y for Yes, the term

is even and N for NO, the term is not even, we have the following pattern commencing with the

first term: NYNYNYNYNNYNYNYNYNNYNYNYNN etc. In any grouping of nine, the Y's

correspond to those term numbers congruent to 2, 4, 6 or 8 modulo 9 while, in contrast, the N's correspond to those term numbers congruent to 0, 1, 3, 5 or 7 modulo 9. An integer is divisible by $2^2 = 4$ if and only if the last two digits of the integer is divisible by 4. Based on the structure of the sequence, only integers which terminate in the digits 76 or 32 are candidates. The corresponding term numbers represent integers congruent to either 4 or 8 modulo 9 and are of the form $9 \cdot k + 4$ *or* $9 \cdot k + 8$. Since an integer is divisible by 8 if and only the last three digits of the integer is divisible by 8, there is only one possibility based on the structure; namely 432. This corresponds to integers congruent to 8 modulo 9. Such integers are of the form $9 \cdot k + 8$. No integer in the sequence is divisible by 16 which is immediate by case analysis. One needs

to examine the last four digits of such integers. Observe that 9876, 7654, 5432, 2198 would constitute the only possibilities for even integers and none is divisible by 16 as is easily checked. Hence no power of two higher than 2^3 can ever occur as a factor of any term in this sequence.

Divisibility by 5 requires one to look at the units digit which must be 0 or 5. There are no zeros in the sequence. Hence a 5 occurs as the ending digit with the fifth term and every ninth term thereafter. Succinctly stated, the corresponding term numbers are congruent to 5 modulo 9 and are of the form 9 * k + 5.

 Let us for simplicity call terms of the sequence such as

$S(9) = 987654321$, $S(18) = 987654321987654321$, etc. which contain all nine digits in descending order complete groupings. It is immediate that any term that forms a complete grouping is divisible by 9; for the sum of the digits of such integers is an integer multiple of 45. Appending the digit 9 to any complete grouping preserves the test for divisibility by 9. This establishes the truth of Conjecture 8. Finally, we verify the truth of Conjecture 9. Divisibility of an integer by 11 requires one to consider the alternating sum of the digits from right to left.

Clearly 987654321 is divisible by 11; for $1-2+3-4+5-6+7-8+9 = 0 \equiv 0 \pmod{11}$.

Appending complete groupings to 987654321 clearly preserves the desired result.

Concluding Remarks

Recursive sequences are appropriate for obtaining new and illuminating insights into dynamic mathematics. Additionally, such sequences illustrate the essence of the nature of mathematics as the science of patterns. This article incorporated discrete mathematical ideas entailing patterns and modular arithmetic with the aid of technology such as MATHEMATICA. The reader is cordially invited to partake further of this and other Fibonacci-like sequences exploring and discovering a world of possibilities in the process.

References

1. Mathematica v. 8.0, *Wolfram Research, Inc.,* Champaign, IL. (2010)

2. *MathWorld-A Wolfram Resource,* Wolfram Research, Inc, Champaign, IL. (2012)

3. The On-Line Encyclopedia of Integer Sequences (njas@research.att.com) (2012)

AN ANALYSIS OF MULTISTATE LIGHTS OUT ON A CUBE

John Antonelli

Crista Arangala

Elon University
Department of Mathematics and Statistics
Elon, NC 27244
ccoles@elon.edu

Abstract

This paper analyzes the multistate Lights Out game on a cube. With the goal of the original lights out game being to get all of the lights off, we explore how to get all buttons from one state to the next on a cube with simultaneous presses. Solutions for an arbitrary number of states is discussed and a gathering technique for general initial conditions is presented.

Introduction

The original Lights Out game made its way into main stream when Tiger Electronics created their hand-held Lights Out Puzzle in 1995. In the original game, each of 25 lights, in a 5 × 5 grid, starts in either the on state or in the off state and the goal of the game is to get all of the lights off (or out) by pressing the buttons. If a button is pressed, it changes its own state and the state of all vertically and horizontally adjacent buttons.

In [7], Joyner discusses the mathematics behind Lights Out. Joyner gives insight into the classic 5×5 Lights Out Puzzle, the 4×4 Lights Out keychain, 6×6 Deluxe Lights Out, Rubik's Clock, and Alien Tiles [9]. These games vary as some of the puzzles change not only grid size or shape, but also the response when buttons are pressed and/or the number of buttons states. A linear algebra approach to a solution to the Lights Out Puzzle was introduced by Anderson and Feil[1], who used nullspaces of matrices to look predominantly at the 5 × 5 classic Lights Out Puzzle. In [4], the authors present several arguments regarding the existence of a solution in

the n × n Lights Out puzzle and use odd dominating sets to discuss the existence of a solution to the Lights Out puzzle on a general graph as well.

Aráujo[2], in 2000, and Missigman and Weida [11], in 2001, offered complete solutions to the Lights Out Puzzle on the 4 × 4 torus. In addition to these solutions of the many versions of the Lights Out Puzzle, researchers have explored the game with different rules for changing the state of the buttons. Buttons are considered inclusive if when a button is pressed it toggles its own state as well as the state of all adjacent buttons. Buttons are known as exclusive when the pressed button does not change states, only the adjacent buttons change states. Sutner[14] and Klostermeyer and Goldwasser [8] named the inclusive rule $\sigma+$, and the exclusive rule simply σ game.

In this paper, we explore the multistate Lights Out game in which the buttons can take on more than 2 states. Many of the arguments related to parity in the "binary" Lights Out Puzzle no longer apply in the multistate Lights Out game. Particular multistate Lights Out games have been explored including the 3-color Lights Out game on the 4 × n grid [3] and the 5-color Lights Out game on the 6 × n grid[5]. In this paper, we focus on the multistate Lights Out game on the cube and describe in detail the cube size and number of states which produce a solution. A gathering technique, similar to that proposed in [1], on the cube is also described.

The Lights Out Game

Lights Out games can be represented by a grid of buttons. In the binary, two state, Lights Out game, buttons start in the on or off state with a goal of finding a set of presses to turn all of the buttons off. While in the multistate games with p colors, each button has an initial state of one of the colors $0, 1, \ldots, p - 1$. The initial state vector $\vec{i}$ represents the initial state of the grid. The final state vector $\vec{f}$ represents the states of the buttons after all presses are performed. The goal of the multistate game is to find a press vector that, when applied to the grid of all lights in state\color 0, toggles each button to state\color 1. Note that each button toggles itself and all of its adjacent buttons. One can see how the multistate aspect of the game complicates the solution in the example presented in Figure 1. In the example in Figure 1, we press the buttons which are a solution to the binary Lights out game, **1, 3, 7, and 9**; however in doing so construct a final state which has buttons in both states 1 and 2.

In the 3 × 3 Lights Out 3 color game in Figure 1, the initial state vector is $\vec{\imath}=(0,0,0,0,0,0,0,0,0)$.

When button 1, 3, 7, and 9 are pressed, the press vector of $\vec{p}=(1,0,1,0,0,0,1,0,1)$ results in a final

state vector of $\vec{f}=(1,2,1,2,0,2,1,2,1)$.

Solutions to the Lights Out game may be found through trial and error, however we will

introduce a linear algebra approach here. The adjacency matrix A is defined such that if buttons

m and n are adjacent then $A_{m,n} = A_{n,m} = 1$, otherwise $A_{m,n} = A_{n,m} = 0$. Also, $A_{m,m} = 1$ since a

pressed button is adjacent with itself. The adjacency matrix associated with the 3 × 3 Lights Out

game is

$$A = \begin{pmatrix} 1 & 1 & 0 & 1 & 0 & 0 & 0 & 0 & 0 \\ 1 & 1 & 1 & 0 & 1 & 0 & 0 & 0 & 0 \\ 0 & 1 & 1 & 0 & 0 & 1 & 0 & 0 & 0 \\ 1 & 0 & 0 & 1 & 1 & 0 & 1 & 0 & 0 \\ 0 & 1 & 0 & 1 & 1 & 1 & 0 & 1 & 0 \\ 0 & 0 & 1 & 0 & 1 & 1 & 0 & 0 & 1 \\ 0 & 0 & 0 & 1 & 0 & 0 & 1 & 1 & 0 \\ 0 & 0 & 0 & 0 & 1 & 0 & 1 & 1 & 1 \\ 0 & 0 & 0 & 0 & 0 & 1 & 0 & 0 & 1 \end{pmatrix}.$$

Figure 2
Button numbering on the cube for the 3x3 face.

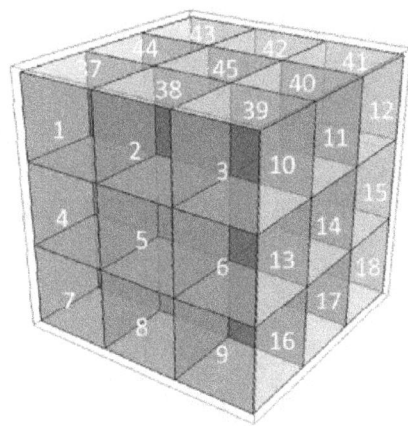

The product $A\vec{p}$ represents the buttons that are toggled and thus the goal of the Lights Out game with p colors is to find $\vec{p}$ such that $A\vec{p} + \vec{0} = \vec{1}\ mod\ p$.

The Lights Out Cube

In this paper, we will focus on the cube with an n×n face, and thus $6n^2$ buttons. An example of how the cube will be numbered in 3 dimensions and flattened in 2 dimensions can be seen in Figures 2 and 3.

The goal of this multistate Lights Out game, with p prime colors, is to start with a cube where all buttons are the same color, color 0, and to find a sequence of presses, that when applied changes all buttons to the next color, color 1. The necessity to work in a field when attempting to find the inverse of the adjacency matrix drives our choice for prime number colors; however when using other techniques no such requirement exists. Some examples of solutions to multistate Lights Out cube games can be seen in Figures 4 and 5.

Figure 3
Example of button numbering in flattened cube with a 3 × 3 face.

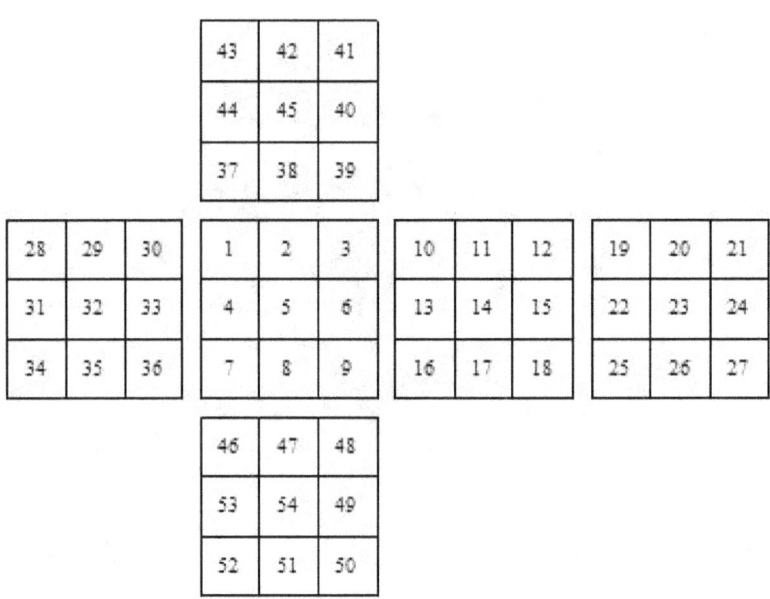

Figure 4
Solution for game with 2 × 2 face with 3 colors

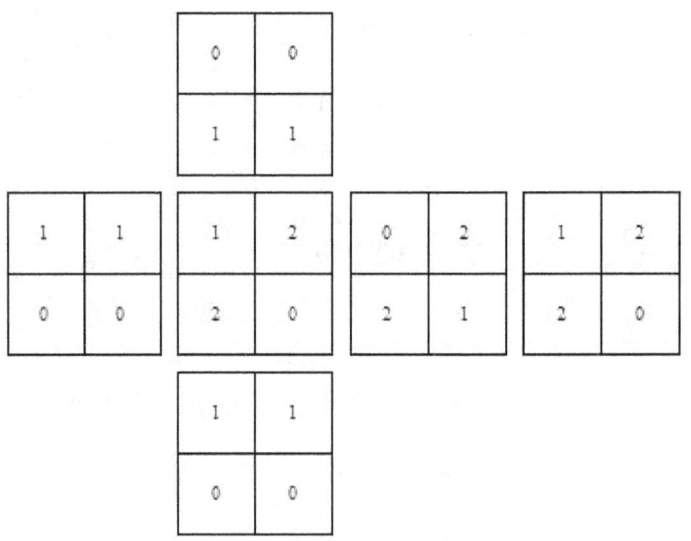

Figure 5
Solution for game with 4 × 4 face with 7 colors

2	1	5	4
1	6	0	5
5	0	6	1
4	5	1	2

2	1	5	4	4	5	1	2	2	1	5	4	4	5	1	2
1	6	0	5	5	0	6	1	1	6	0	5	5	0	6	1
5	0	6	1	1	6	0	5	5	0	6	1	1	6	0	5
4	5	1	2	2	1	5	4	4	5	1	2	2	1	5	4

2	1	5	4
1	6	0	5
5	0	6	1
4	5	1	2

When $p \neq 5$ in a cube with an n×n face, for any natural number n. In an n×n face cube, all buttons are adjacent to 5 buttons, including themselves. So if we take a cube and start with all buttons in one color, color 0, we can toggle all buttons to the next color, color 1 if p is not 5.

Theorem 1. Given a n × n face cube there is a sequence of presses that solve the multistate Lights Out cube game with p colors for all prime $p \neq 5$, n $\in$ **Z**⁺.

Proof. Let A be the $6n^2 \times 6n^2$ adjacency matrix for our $n \times n$ face cube. We want to find a $6n^2 \times 1$ press vector $\vec{p}$ such that $A\vec{p} = (1,...,1)^T$. Toward that end, let $\vec{x} = (x,...,x)^T$ for some variable x. Then we know that $A\vec{x} = 5\vec{x}$, because every button when pressed is adjacent to 5 buttons including itself. So we want to find a positive integer x such that $A\vec{x} = (1,...,1)^T$. Equivalently this also means that, $5\vec{x} = (1,...,1)^T \bmod p$, or $5x \equiv 1 \bmod p$. Since $p \neq 5$, the above congruence has a unique solution mod p, by standard result in elementary number theory. We will call that solution s. Then letting $\vec{p} = (s,...,s)^T$, we have:

$\vec{p} = A(s,\ldots,s)^T = (5s,\ldots,5s)^T = (1,\ldots,1)^T mod\ p.$

This proof shows that it is possible to go from a solid color(color 0), and find a press vector that will turn the cube into color 1, when $p \neq 5$ in any n × n face cube.

When $p = 5$ **in a cube with a** n × n **face, for any natural number** n**.** The solution is given for $n = 5$ and $p = 5$ in Figure 6. One might notice in the solution presented in Figure 6, that the presses on each face are the same. Denote the presses for this single face by X. X can also be seen in Figure 6.

Figure 6
Solution for Lights Out cube game with 5 × 5 face and 5colors (L) and *X*, the presses on a single side for this solution (R).

1	3	2	0	1
0	0	4	0	3
2	4	0	4	2
3	0	4	0	0
1	0	2	3	1

1	3	2	0	1
0	0	4	0	3
2	4	0	4	2
3	0	4	0	0
1	0	2	3	1

1	3	2	0	1
0	0	4	0	3
2	4	0	4	2
3	0	4	0	0
1	0	2	3	1

1	3	2	0	1
0	0	4	0	3
2	4	0	4	2
3	0	4	0	0
1	0	2	3	1

1	3	2	0	1
0	0	4	0	3
2	4	0	4	2
3	0	4	0	0
1	0	2	3	1

1	3	2	0	1
0	0	4	0	3
2	4	0	4	2
3	0	4	0	0
1	0	2	3	1

1	3	2	0	1
0	0	4	0	3
2	4	0	4	2
3	0	4	0	0
1	0	2	3	1

If $p = 5$ and $n = 5k$ where k is an integer, then a solution for this multistate Lights Out cube game with 5 colors can be found by strategically repeating X, k times, as in Figure 7.

Figure 7

A solution to Lights Out game with 5k × 5k face and 5 colors

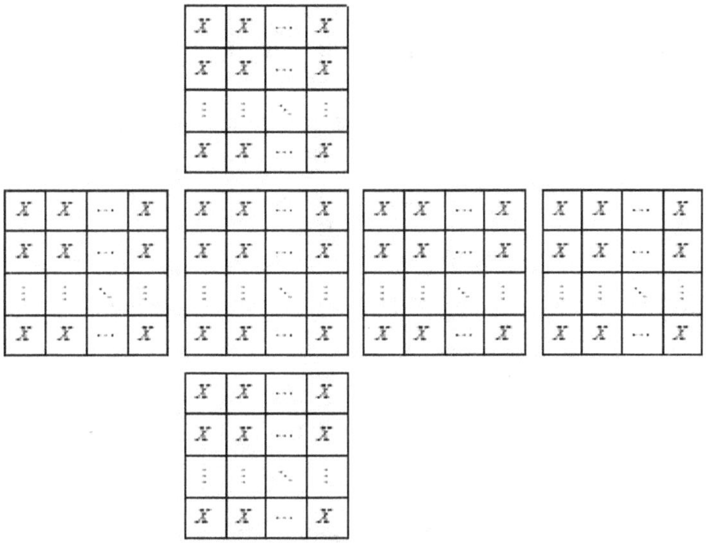

If $p = 5$ and n is not a multiple of 5, we see that the affiliated adjacency matrix is singular and thus we wish to determine if there exists a press vector, $\vec{p}$ such that $A\vec{p} = \vec{1}$. This is equivalent to determining if $\vec{1}$ is in the orthogonal complement to the nullspace of A.

When $p = 5$, pressing every button is equivalent to not pressing them at all, and thus $\vec{1}$ is in the nullspace of A. Because $\vec{1} \cdot \vec{1} \equiv 0 \bmod 5$ when n is not a multiple of 5, no solution exists for the multistate Lights Out cube game with 5 colors and a $n \times n$ face, $n \neq 0 \bmod 5$.

General Solutions To The Cube Game

In the previous sections, we explored only those multistate Lights Out cube games with initial state $\vec{0}$ and final state $\vec{1}$. Here we explore games with other nonuniform final state vectors. We begin with a particular example. In the Lights Out cube game with a 3×3 face and 5 colors, we wish to determine if there a solution if $\vec{i} = \vec{0}$ and $\vec{f}$ is defined such that opposite sides of the cube are the same colors. That is the final state has all buttons on sides 1 and 3 of color x, all buttons on sides 2 and 4 of color y, and all buttons on top and bottom of color z where x, y, and z are distinct. If this final state is in the range then it is in the orthogonal complement to the nullspace and thus is orthogonal to each nullvector of A. Taking the dot product of $\vec{f}$ and each nullvector of A results in the following equations which must hold true in order for a solution to exist.

$$18x + 18y + 18z \equiv 0 \bmod 5,$$
$$20x + 20y + 20z \equiv 0 \bmod 5,$$
$$20x + 20z \equiv 0 \bmod 5,$$
$$20x + 20y \equiv 0 \bmod 5.$$

All of the above equations hold true for any choice of x, y, and z except for the first. Therefore the above multistate Lights Out game has solutions when the opposite side color states are $(0, 2, 3)$ or $(0, 1, 4)$.

We next wish to look at which 3×3 multistate Lights Out games have a solution with $\vec{i} = \vec{0}$ and $\vec{f}$ defined such that each side of the cube is a uniform color and distinct. Specifically, the final state has all buttons on side i of color $i - 1$, $1 \leq i \leq 6$, and $p > 5$.

Theorem 2. Given the multistate Lights Out game on a cube with a 3×3 face and $p > 5$ prime, there exists a press vector, $\vec{p}$, such that $A\vec{p} + \vec{i} = \vec{f}$, $\vec{i} = \vec{0}$ and $\vec{f} = (0, 0, \ldots, 0, 1, 1, \ldots, 1, 2, 2, \ldots$ $, 2, 3, 3, \ldots, 3, 4, 4, \ldots, 4, 5, 5, \ldots, 5)$ *where each of the numbers in $\vec{f}$ are repeated 9 times.*

Proof. For the cube with the 3×3 face and $p > 5$ prime, the adjacency matrix A has nullity 3. Denote the sequence $0, 1, 0, p-1$ as w_1 and the sequence $p-1, 0, 1, 0$ as w_2, the sequence $0, p-1, 0, 1$ as w_3, sequence $1, 0, p-1, 0$ as w_4, and w_5 is the sequence of 9 zeros. The nullvectors of A are,

$$v1 = (w1, 0, w2, 0, w2, w1, 0, w1, w2, 0, w2, w1, 0, w2, w2, 0, w2, w2)$$
$$v2 = (w2, 0, w1, w5, w2, 0, w1, w5, w4, w4, 0, w2, w2, 0)$$
$$v3 = (w4, 0, w3, w4, 0, w3, w4, 0, w3, w4, 0, w3, w5, w5)$$

Since, $\vec{f} \cdot v1 = 30 \cdot p \equiv 0 \bmod p$, $\vec{f} \cdot v2 = 22 \cdot p \equiv 0 \bmod p$, and $\vec{f} \cdot v3 = 12 \cdot p \equiv 0 \bmod p$.

Thus, $\vec{f}$ is in the orthogonal complement to the nullspace of A and there is a sequence of pushes that will take the game from buttons in all color 0 to the final state $\vec{v}$. In the next section, we explore a technique for determining other final vectors that have a solution in the multistate Lights Out cube game with a 3×3 face.

Gathering The Cube

In [1], the authors present a gathering technique for determining which initial states for the 3×3 binary grid game have solutions. We will present a similar technique for determining which initial states have solution in the cube game with a 3×3 face and 5 colors. We begin with an example of this gathering technique on the initial state pictured in Figure 8.

Figure 8

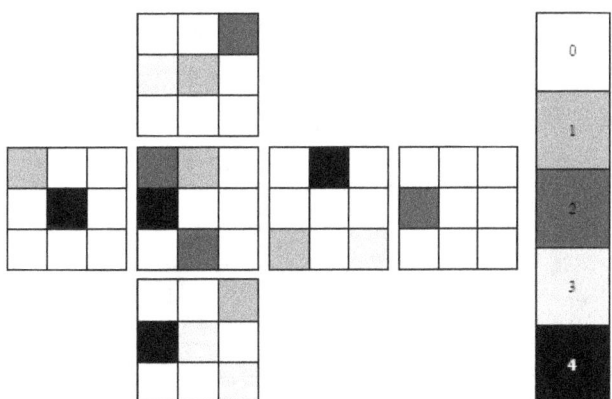

The goal in the gathering technique is to work down the grid making each consecutive row of buttons zero by pushing the buttons directly below them an appropriate number of times. Note that under this gathering technique, buttons 10, 11, 12, 19, 20, 21, 28, 29, and 30, are not pressed. Thus given an initial vector this gathering technique will only determine if there is a solution without pushing these buttons. In the grid gathering, the final row of buttons, buttons 16, 17, 18, 25, 26, 27, 34, 35, 36, 50, 51, and 52, which are not necessarily all zeros are then put into equivalence class, which will be discussed later. We will begin this gathering at the top row in

the flattened version of the cube and work of way down.

Figure 9
Gathering Steps 1 through 4.

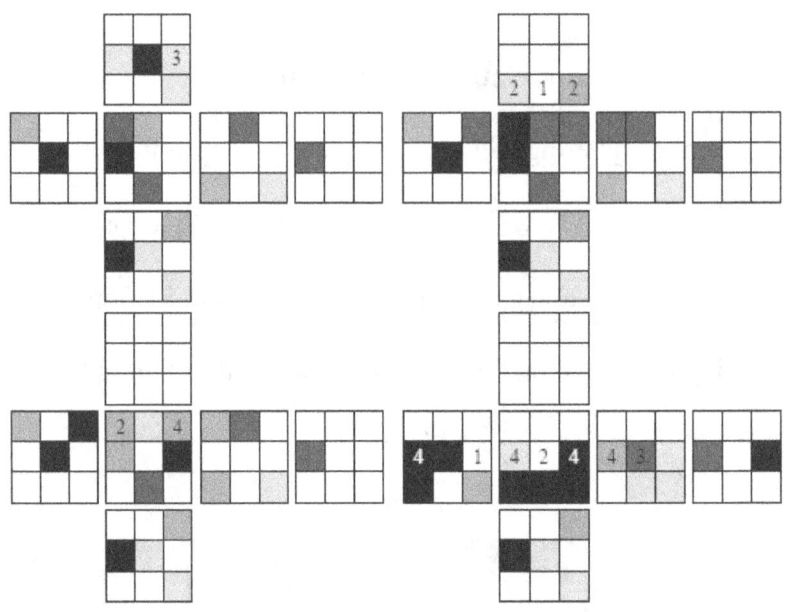

Figure 10
Gathering Steps 5 through 8

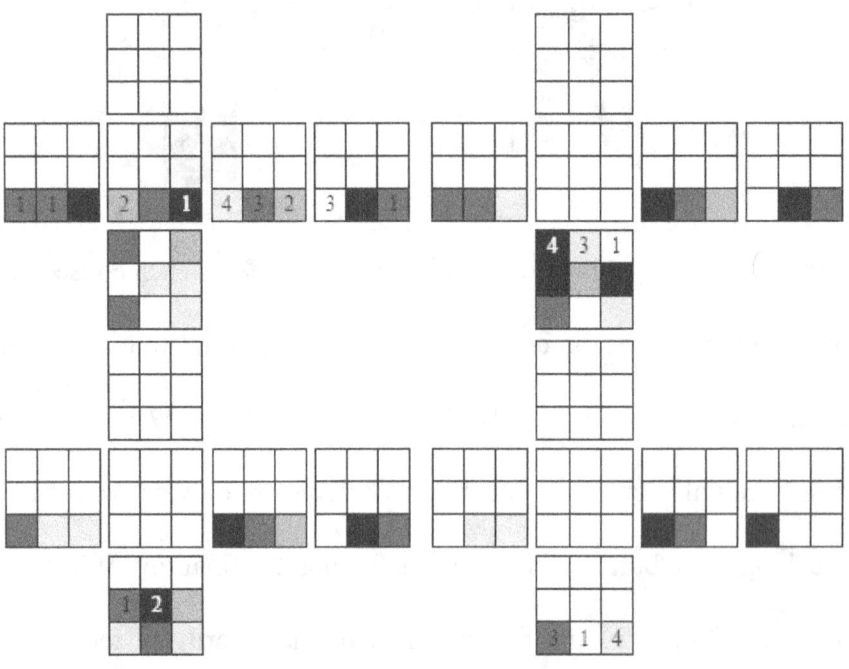

In order to decide if this gathering produces a solution, we must gather all boards with buttons of color 0 except buttons 41, 42, 43. If the additive inverse of

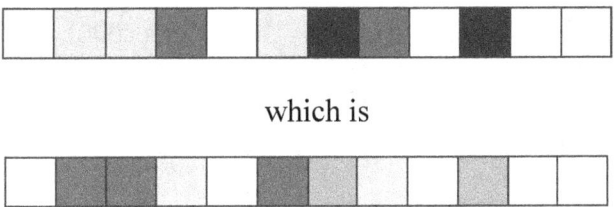

which is

results from gathering the buttons 43 = *x*, 42 = *y*, and 41 = *z* then the solution to the the original configuration is to press button 43, *x* times, 42, *y* times, and 41, *z* times, and then gather the board to toggle all buttons to color 0. The results of the gathering of all boards with buttons of color 0 except buttons 41, 42, 43, which we call the equivalence classes, can be found in the appendix. Since no equivalence class results in the gathering in Figure 11, no solution exists (where buttons 10, 11, 12, 19, 20, 21, 28, 29, and 30 are not pressed) for the initial condition given.

Further Questions

In this paper, the main result presented relates to solutions to the multistate Lights Out cube game with initial state $\bar{0}$ and final state $\bar{1}$. We explored a variety of other questions related to multistate Lights Out cube games. There are still many open questions related to general solutions to the multistate Lights Out cube game.

References

1. M. Anderson and T. Feil, "Turning Lights Out with Linear Algebra", *Mathematics Magazine* 71(4), (1998), 300-303.

2. P. Araujo, "How to turn all the lights out", *Elem. Math.* 55, (2000), 135-141.

3. C. Arangala, "The 4xn Multistate Lights Out Game", *Mathematical Sciences International Research Journal*, Vol 1, Number 1, 10-13.

4. C. Arangala, J. T. Lee, B. Yoho, "Turning Lights Out", *UMAP/ILAP/BioMath Modules 2010: Tools for Teaching*, edited by Paul J. Campbell. Bedford, MA: COMAP, Inc., (2010), 1-26.

5. C. Arangala and M. MacDonald, "The 6*xn* Five Color Lights Out Game", *Journal of Recreational Mathematics,* Vol. 38(1), 38-44, 2014.

6. M. Hunziker, A. Machiavelo, J. Park, "Chebyshev Polynomials over Finite Fields and Reversibility of Automata on Square Grids", *Theoretical Computer Science.* 320(2-3), (2004), 465-483.

7. D. Joyner, *Adventures in Group Theory: Rubik's Cube, Merlin's Machine, and Other Mathematical Toys*, The Johns Hopkins University Press, 2002.

8. J. Goldwasser and W. Klostermeyer, "Maximum orbit weight in the sigma-game and lit-only sigma-game on grids and graphs", *Graphs and Combinatorics*, 25 (2009), 309-326.

9. P. Maier and W. Nickel, "Attainable Patterns in Alien Tiles", *American Mathematical Monthly* 114(1),(2007), 1-13.

10. O. Sanchez and C. Flores, "Two Reflected Analysis of Lights Out", *Mathematics Magazine*, 74(1),(2001) 295-304.

11. J. Missigman and R. Weida, "An Easy Solution to Mini Lights Out", *Mathematics Magazine*, 74(1), (2001), 57-59.

12. O. Sanchez and C. Flores, "Two Reflected Analysis of Lights Out"', *Mathematics Magazine*, 74(1), (2001),295-304.

13. K. Sutner, "Automata and Chebyshev Polynomials", *Theoretical Computer Science*, 230, (2000), 49-73.

14. K. Sutner, "Linear cellular automata and the Garden-of-Eden", *Math. Intelligencer* 11 (1989), 49-53.

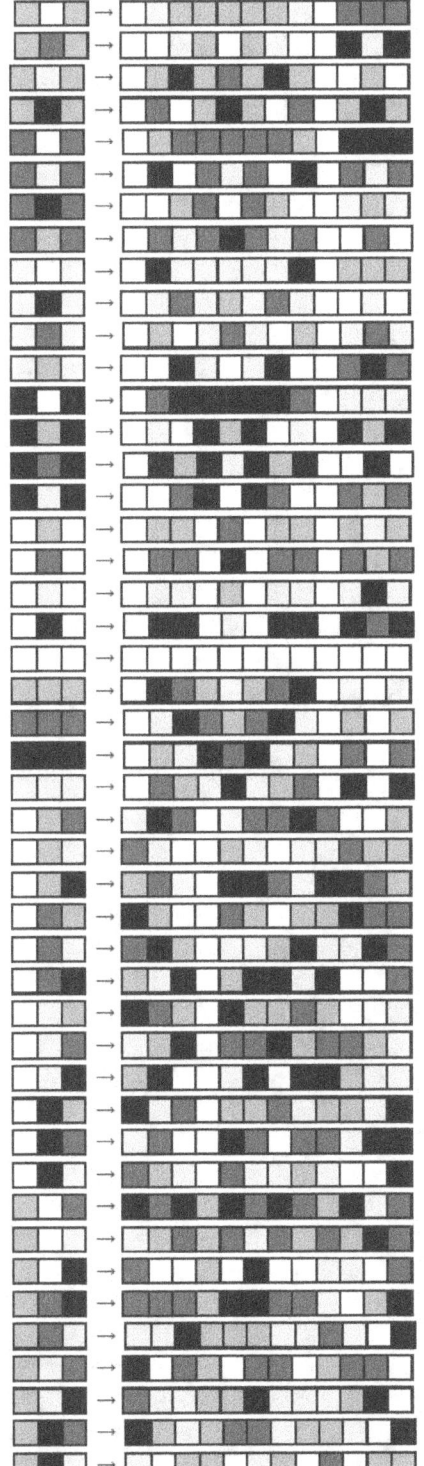

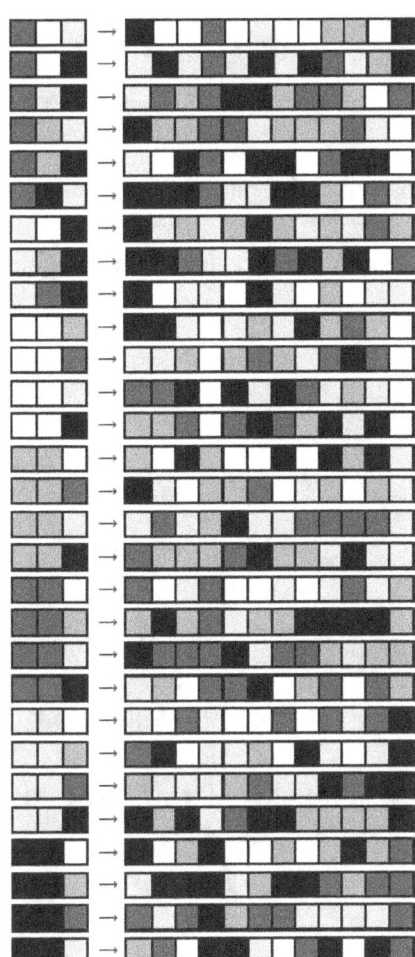

A GRAPHICAL SOLUTION TO THE MONTMORT MATCHING PROBLEM

Diego Castano

Nova Southeastern University

Division of Mathematics, Science, and Technology

3301 College Ave.

Fort Lauderdale, FL 33314

castanod@nova.edu

Abstract

 A graphical, or diagrammatic, method not relying on any formal combinatorial tools is used to solve a variation of the card matching problem. In the card problem, one is interested in determining the probability of matching cards from two shuffled decks on simultaneous draws.

Suppose there are N envelopes addressed to N distinct people, and N distinct letters addressing each of those people. The problem under consideration is to determine the probability of randomly stuffing some, or all, of the envelopes incorrectly. This is a variation of the Montmort card matching problem [1]. That problem was solved by Montmort (and Bernoulli) and subsequently generalized and solved by others, including Euler and Laplace (see [3] and references therein). More recently, generalizations and solutions can be found in the works of [2, [3], [4], and [5].

The method employed here is graphical in that a connection is drawn between abstract combinatorial symbols and more intuitive graphs.

Figure 1: An N-pair graph

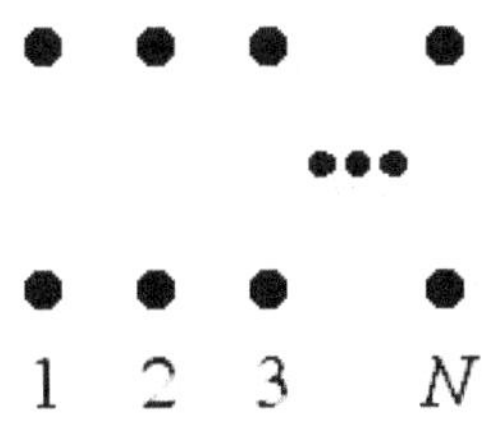

For example, Figure 1 represents an unconnected graph of N related pairs of points. A related point pair will refer to the two points directly above and below each other in a graph. The numbers displayed in the figure simply count the pairs; they do not identify the pairs. The ordering of pairs is unimportant. A connected graph is one with connections between unrelated points (*i.e.*, upper points are connected with non-related lower points). Figure 2 shows a connected 3-pair graph.

Figure 2: A connected 3-pair graph

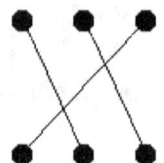

A boxed N-pair graph will represent the total number of connected N-pair graphs, subject to possible constraining connections. Figure 3 shows the counting represented by a box with no constraining connections, which will be labeled E_N. This box can be expressed in terms of other, constrained boxes. The boxes in figure 3 with the one constraining connection represent the number of graphs with this type of connection. These boxes will be labeled Q_{N-1}.

Because the order of the pairs in the graphs is unimportant, it is clear that all the boxes with one constraining connection count the same number of graphs (Fig. 4).

Figure 3: The calculus of N-pair boxes

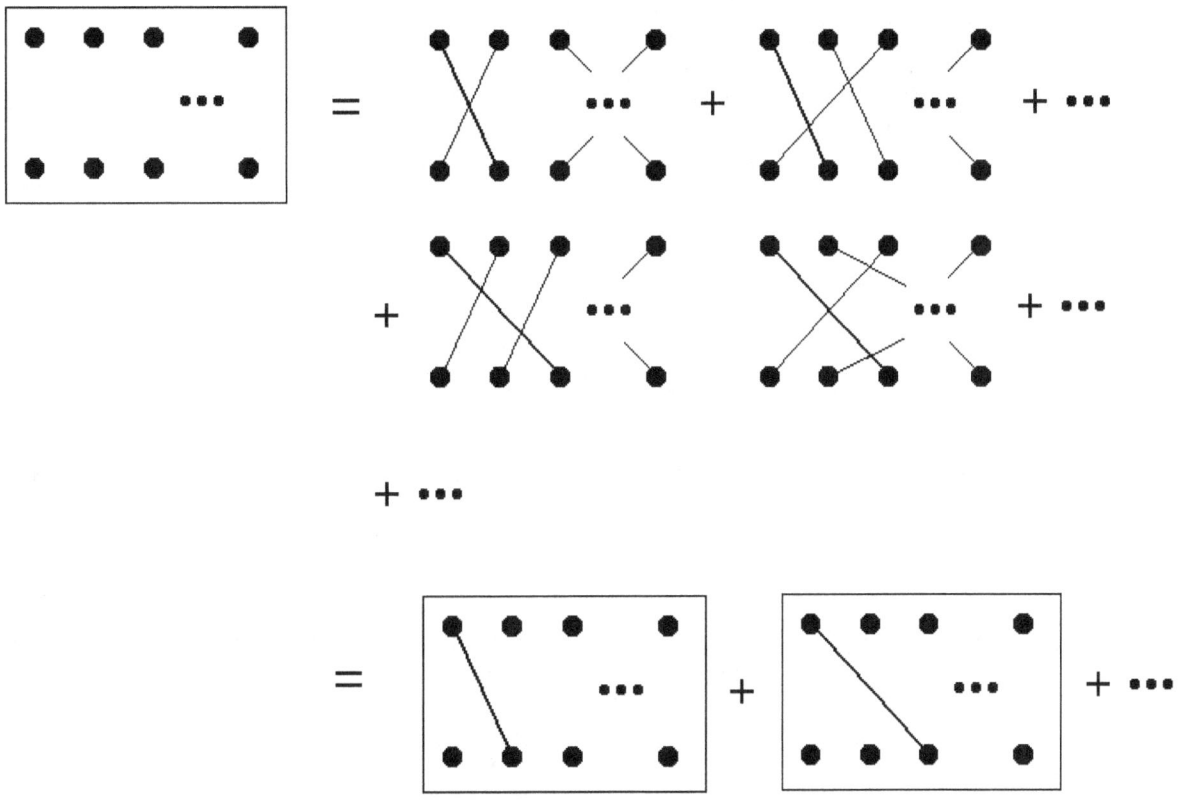

Figure 4: Classification of connected graphs

Formally, the calculus of Figs. 3 and 4 can be expressed as

$$E_N = (N-1)Q_{N-1}. \tag{1}$$

Note that any crossed constrained box, such as in Fig. 5, is equivalent to an unconstrained box with two fewer pairs.

Figure 5: A crossed constrained box

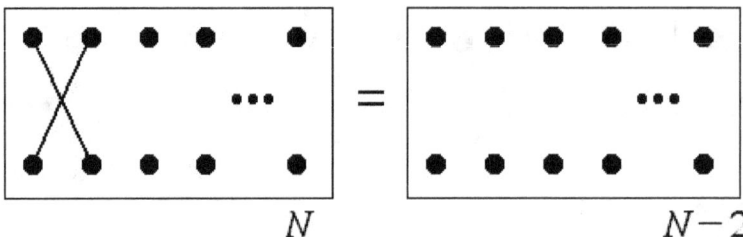

$$N \qquad\qquad N-2$$

Figure 6: Difference relation between constrained boxes

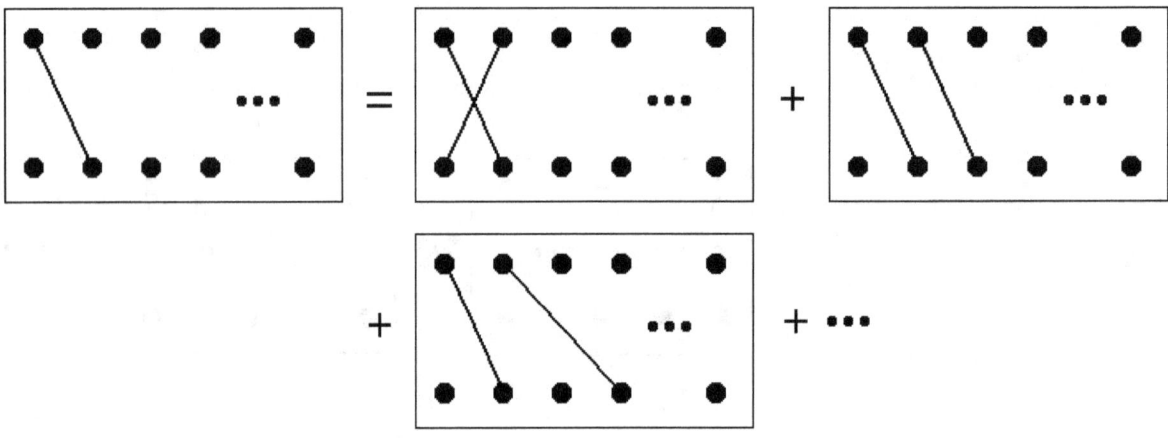

Using this fact and motivated by Fig. 6, it follows that

$$Q_{N-1} = E_{N-2} + (N-2)Q_{N-2}. \qquad (2)$$

Solving for Q_N from Eq. (1) and inserting into (2) yields the following recursion relation

$$E_{N+1} = N(E_N + E_{N-1}) \qquad (3)$$

It is evident that

$$E_1 = 0, \qquad (4a)$$

$$E_2 = 1. \qquad (4b)$$

Consistency with Eq. (3) then requires

64

$$E_0 = 1. \tag{5}$$

Note that recursion relation (3) is satisfied by the factorial function

$$\Pi(N) = N! \tag{6}$$

However, the sequence generated by Eq. (3) with conditions (4a)-(5) is

$$
\begin{array}{ccccccccc}
N & 0 & 1 & 2 & 3 & 4 & 5 & 6 & 7 \\
E_N & 1 & 0 & 1 & 2 & 9 & 44 & 265 & 1854
\end{array}
\cdots
$$

Inspection of this sequence leads to another recursion relation

$$E_N = NE_{N-1} + (-1)^N. \tag{7}$$

Iterating this relation $N - m - 1$ times yields

$$
\begin{aligned}
E_N &= N(N-1)(N-2)\cdots(m+1)E_m \\
&+ N!\left[\frac{1}{N!} - \frac{1}{(N-1)!} + \frac{1}{(N-2)!} - \cdots + \frac{(-1)^{N-m-1}}{(m+1)!}\right](-1)^N.
\end{aligned}
\tag{8}
$$

The second term can be written

$$
N!\left[\sum_{n=m+1}^{N}\frac{(-1)^{N-n}}{n!}\right](-1)^N = N!\sum_{n=m+1}^{N}\frac{(-1)^n}{n!}
\tag{9}
$$

By taking $m = 1$, Eq. (9) reduces to the simple form

$$E_N = N!\sum_{n=2}^{N}\frac{(-1)^n}{n!}. \tag{10}$$

Figure 7: Two examples of generalized N-pair graphs

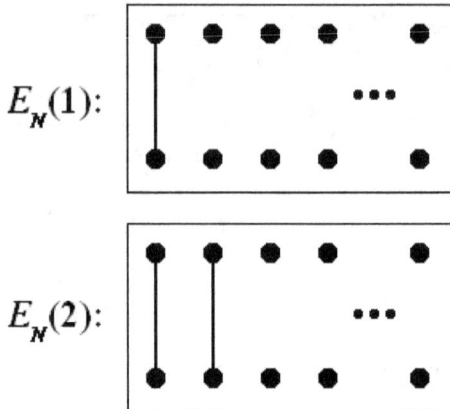

So far the analysis has assumed N-pair graphs with no related pair connections. The results can be generalized to graphs with related pair connections. Assuming N-pair graphs with k related pair connections, the number of ways to connect the remaining points in a non-related way can be deduced from inspection of Fig. 7. The result of Eq. (10) implicitly had $k = 0$. For general k, the result is

$$E_N(k) = \binom{N}{k} E_{N-k}(0).$$
(11)

These results can now be used to determine the probability of stuffing N envelopes with letters so that k envelopes will be stuffed appropriately. The total number of ways to stuff the N envelopes is $N!$, therefore

$$P_N(k) = \frac{E_N(k)}{N!} = \frac{1}{k!} \sum_{n=2}^{N-k} \frac{(-1)^n}{n!}.$$
(12)

It is interesting to note that in the limit that $N \to \infty$, the probability of stuffing all envelopes incorrectly is

$$P_\infty(0) = e^{-1}.$$

(13)

References

1. P. R. de Montmort, *Essay d'Analyse sur les Jeux de Hazard*, Jacque Quillau, Paris, 1708.

2. Don Rawlings, The Poisson Variation of Montmort's Matching Problem, *Mathematics Magazine*, **73**, No. 3 (2000) 232-234.

3. Gabriela R. Sanchis, Swapping Hats: A Generalization of Montmort's Problem, *Mathematics Magazine*, **71**, No. 1 (1998) 53-57.

4. N. S. Mendelsohn, Symbolic Solution of Card Matching Problems, *Bull. Amer. Math. Soc.*, **52** (1946) 918-924.

5. Irving Kaplansky, Symbolic Solution of Certain Problems in Permutations, *Bull. Amer. Math. Soc.*, **50** (1944) 906-914.

BEHFOROOZ CALENDARICAL MAGIC SQUARES

Hossein Behforooz

Mathematics Department

Utica College

Utica, NY 13502

hbehforooz@utica.edu

www.utica.edu/hbehforooz

Abstract

This is another interesting fun topic in recreational mathematic. Simply, select any month in any desktop or wall calendar and draw any three by three or four by four table and rearrange the date numbers in these tables to create different magic squares. Yes MATH is fun and MATH is cool.

Introduction

Martin Gardner has a simple but interesting puzzle in his book "Mathematics, Magic and Mystery", page 48 [1]. It says, ask the audience in a party to select any three by three table from any month of a wall calendar and ask them to give you the smallest number. You can tell them immediately the sum of all nine numbers in that table. The trick is that by adding eight to that smallest number you will get the middle number and the total of those nine numbers is equal to nine times the middle number. There is another puzzle similar to this in "World's Most Baffling Puzzles", from Charles Barry Townsend, page 68 [2]. When I saw these two problems, I became curious to investigate and see if we can make any magic squares from calendar dates. Here is the result of this research.

Puzzle I: Arrange the dates of any 3 by 3 table on any wall calendar to make a magic square.

Solution: Last year, October 15th was my 70th birthday. Let's consider the following three by three table with my birthday at the center.

7	8	9
14	15	16
21	22	23

If we want to create a magic square with these nine numbers, the magic sum will be one third of the total of these nine numbers which is S=45. We cannot move 15 from the middle cell. As you see, the four sets of numbers 7, 15, 23 and 9, 15, 21 and 8, 15, 22 and 14, 15, 16 add up to 45. That means that more than half of the construction is over. We just need to arrange 7, 9, 21 and 23 to complete the magic square. With a simple trial and error method we can see that there is only one solution to this puzzle (regardless of rotation or symmetry) and it is the following three by three magic square with magic sum equal to S=45.

14	9	22
23	15	7
8	21	16

S=45

Interestingly, these numbers satisfy in the following relations, where T_n is the nth triangular number, see also [3].

$$T_{14} + T_9 + T_{22} = T_8 + T_{21} + T_{16} \qquad T_{14} + T_{23} + T_8 = T_{22} + T_7 + T_{16}$$

Of course this technique works for any three by three table on any wall calendar. Also, for any m and d, we can construct a magic square from 9 numbers $m + id + j$; $i, j = 0, 1, 2$, with magic sum equal to S=3m+3d+3.

m	$m+1$	$m+2$
$m + d$	$m + d + 1$	$m + d + 2$
$m + 2d$	$m + 2d + 1$	$m + 2d + 2$

$\Longrightarrow$

$m + d$	$m + 2$	$m + 2d + 1$
$m + 2d + 2$	$m + d + 1$	m
$m + 1$	$m + 2d$	$m + d + 2$

$S = 3m + 3d + 3$

Puzzle II: Arrange the dates of any four by four table on any wall calendar and make a four by four magic square.

Solution: Consider this table from a calendar (you can select any four by four table) with entries in four different colors.

Table 1

1	2	3	4
8	9	10	11
15	16	17	18
22	23	24	25

The sum of these 16 numbers is 208. So, any magic square with these 16 numbers will have a magic sum equal to S=208/4=52. By using the pattern of the Behforooz four by four Latin-Sudoku-Magic Squares (see [4]), we obtain the following four by four magic square. We have colored the entries in four colors to show that the resulting magic square has an orthogonal double diagonal Latin Square pattern.

1	2	3	4
8	9	10	11
15	16	17	18
22	23	24	25

1	16	24	11
25	10	2	15
9	22	18	3
17	4	8	23

Magic square with S=52

In this magic square, besides the rows, columns and diagonals, the magic sum S=52, appears everywhere and these are just few examples:

1+11+17+23=52, 10+2+22+18=52, 1+16+25+10=52, 16 +24+10+2=52, 1+11+25+15=52,

25+15+9+3=52, 16+24+4+8=52, 1+24+9+18=52, 16+11+22+3=52, 1+25+18+8=52,

16+10+3+23=52, 1+25+22+4=52, 16+25+3+8=52, 16+10+9+17=52, 1+25+3+23=52.

There are some relations between the squares of these entries too:

$$1^2 +16^2 +24^2 +11^2 +17^2 +4^2 +8^2 +23^2 = 25^2 +10^2 +2^2 +15^2 +9^2 +22^2 +18^2 +3^2 = 1852$$

$$1^2 +25^2 +9^2 +17^2 +16^2 +10^2 +22^2 +4^2 = 24^2 +2^2 +18^2 +8^2 +11^2 +15^2 +3^2 +23^2 = 1852$$

The diagonal entries and non-diagonal entries have amazing relations. These relations are similar to the relations that we have for the Dürer magic square, (see [5] and [6]).

1+10+18+23+11+2+22+17=16+24+15+3+8+4+9+25,

$1^2 +10^2 +18^2 +23^2 +11^2 +2^2 +22^2 +17^2 = 16^2 +24^2 +15^2 +3^2 +8^2 +4^2 +9^2 +25^2=1852.$

Aren't these relations NEAT stuff? Remember that, for centuries, everybody has had access to these numbers on their wall or desktop calendars but it seems that nobody has noticed the existence of these interesting and amazing magic squares.

Since the sum of four two by two corner squares have the same sum S=52, the above four by four magic square is a gnomon magic square, see [7, page 24] and [10, page 137] .

Here comes my last magic show about our new magic square. There is another way to predict S=52 right at the beginning of this process. Select any number in Table 1, say **16** and delete all entries in the same row and same the column except **16**. Now select another number, say **11** and delete the other numbers, and repeat this and select one more number, say **3** and delete the other numbers. The sum of these four selected and remaining numbers is **S=16+11+3+22 = 52**. This works for any different selection of four numbers in Table 1. This is another neat property.

1	2	3	4
8	9	10	11
15	**16**	17	18
22	23	24	25

1		3	4
8		10	**11**
	16		
22		24	25

1		3	
			11
	16		
22		24	

		3	
			11
	16		
22			

Again, the above method works for any table with 16 numbers *a + nk, b + nk, c + nk, d + nk,* with *n=0, 1, 2, 3*, and we obtain a four by four magic square with magic sum *S=a+b+c+d+6k.*

a	*b*	*c*	*d*
a + k	*b + k*	*c + k*	*d + k*
a+2k	*b+2k*	*c+2k*	*d+2k*
a+3k	*b+3k*	*c+3k*	*d+3k*

a	*b+2k*	*c+3k*	*d+ k*
d +3k	*c + k*	*b*	*a+2 k*
b +k	*a+3k*	*d+2k*	*c*
c+2k	*d*	*a +k*	*b+3k*

S=a+b+c+d+6k

Generalization: Getting the above idea from a calendar we can now generalize this idea to any *n* by *n* table with entries

m, m+1, m+2, ……..,m+(n 1)

m+d, m+d+1, ……,m+d+(n 1)

m+2d, m+2d+1, ……,m+2d+ (n 1)

…………………………………………………..

m+(n 1)d, m+(n 1)d+1, ……., m+(n 1)d+(n 1)

To construct any *n* by *n* magic square with these numbers we must follow the pattern of an orthogonal double diagonal Latin Square of order $n \geq 4$ which is available in the literature. The study of Latin Squares by Euler guarantees the existence of these kinds of orthogonal double diagonal Latin Squares for $n \geq 4$. As Laplace has mentioned, Euler is the master of ALL of US. For more information and the algorithm of construction of orthogonal double diagonal Latin Squares see for example [8]. The magic sum of this magic square with the above numbers is:

$$S = mn + \frac{n(n-1)(d+1)}{2}.$$

With $m = 1, n = 5, d = 10$ we obtain the following five by five magic square with magic sum S=115. Here again 115 is everywhere, for example, the numbers at the center and four corner cells add up to 115 and any five numbers on any + or × shapes add up to 115. This is true if we make a cylindrical magic square from this flat square.

1+33+25+12+44=115, 42+34+25+11+3=115, 24+35+43+1+12=115, 2+15+43+21+34=115,

24+13+2+45+31=115, 2+14+21+43+35=115, 4+33+22+11+45=115, 4+11+23+45+32=115

1+24+45+13+32=115, 4+22+43+11+35=115. 42+24+13+2+34=115, 4+31+25+14+41=115

1	24	42	15	33
45	13	31	4	22
34	2	25	43	11
23	41	14	32	5
12	35	3	21	44

S=115

Dürer or not Dürer that is the question: As you may know, in 1514, the German renaissance painter and graphic artist *Albrecht Dürer* (Fig. 1) included a four by four magic square at the top right corner of his famous engraving entitled *Melencolia* I (Fig. 2) see also [9] and [10]. This piece of art made this magic square very popular in Western countries. Since then this magic square has been called the *Dürer Magic Square* (Fig. 3).

Fig. 1 Albrecht Durer, Self Portrait   Fig. 2 Melencholia I (1514) Engraving

Fig. 3 Yang Hui Dürer Magic Square

Last year (2014), on the celebration of the 500th birthday of the *Melencholia I Engraving* and the popularity of this four by four interesting magic square, there were worldwide presentations, lectures and different events in different museums and universities, (for more information, you can search the internet for these events and presentations). As I have mentioned these historical notes in [11], the fact is that this magic square is not from Albrecht Dürer. It is from China and composed by Yang Hui (Yes, it is made in China), see [12, page 528]. So, it is better to call this famous magic square "Yang Hui Dürer magic square". I have seen this magic square in Iranian literature which is more than 900 years old, see [13], [14]. In [13], the author claims that the Yang Hui Dürer magic square is useful to get rid of the pimples from your face or body (kind of shingle vaccine). This is one example of thousands of different strange and bizarre beliefs about the super powers of the magic squares that we can see in different cultures and different places of the world. Another historical note about this magic square is that in 1209, Yang Hui demanded Genghis Khan, the Emperor of Mongolia, to complete the above magic square, when only few numbers were written in few cells, for more information on this note, see [15]. These are evidences that the above magic square is not from Germany and it is not 500 years old. But Melencholia I Engraving is 500 years old, see also [16].

To continue our discussion, we can use the Yang Hui Dürer magic square pattern to transform or change Table 1 to a new four by four magic square.

1	2	3	4
8	9	10	11
15	16	17	18
22	23	24	25

Table 1

Yang Hui Dürer magic square

25	3	2	22
8	16	17	11
15	9	10	18
4	24	23	1

New Magic Square with S=52

This magic square is also a gnomon magic square and has the same kinds of properties that we have mentioned above for our four by four magic square. Here, for example, diagonal entries verses non-diagonal entries satisfy in the following three relations for $n = 1, 2, 3,$

$$25^n +16^n +10^n +1^n +22^n +17^n +9^n +4^n = 3^n +2^n +11^n +18^n +23^n +24^n +15^n +8^n$$

Final Thought: I have read this somewhere. Somebody has asked the Italian renaissance sculpture, painter, architect, and poet Michelangelo about his masterpiece the **statue of David** (Fig. 4). He asked how Michelangelo has made that beautiful and lovely statue of David from a big piece of hard rock marble and Michelangelo simply replied that he did not make it at all, because it was there inside that big rock and he has just chopped and cleaned the rock to bring the statue out. That is all he said. For many years the date numbers of calendars were in front of all of US on our desks, walls or in our pockets and we did not see or notice that there are many beautiful magic squares inside these numbers. I have just rearranged these numbers and made the above magic squares. That is all. Do Math and enjoy Math and remember that Math is Fun and Cool. (I have copied all of these figures from different sections of www.wikipedia.com).

Fig. 4 The **Statue of David,**
Completed by Michelangelo in 1501-1504

References

1. M. Gardner: *Mathematics, Magic and Mystery,* BN Publishing, 2012,

   (see, www.nbpublishing.com).

2. C. B. Townsend, *World's Most Baffling Puzzles*, Sterling Publishing Company, New York, 1991.

3. O. O'Shea, Some Words on the Lo Shu, *Journal of Recreational Mathematics*, vol. 35 (1) 23-29, 2006.

4. H. Behforooz, Behforooz Magic Squares Derived from Magic-Latin-Sudoku Squares, *Journal of Recreational Mathematics*, vol. 36 (4) 287-293, 2007.

5. H. Behforooz, Permutation Free Magic Squares, *Journal of Recreational Mathematics*, vol. 33 (2) 103-106, 2005.

6. P. D. Schumer: *Mathematical Journeys*, John Wiley & sons, Hoboken, New Jersey, 2004.

7. J. Hunter and J. Madachy, *Mathematical Diversions.* New York: Dover, 1975.

8. J. Straight, Notes on Magic Squares, Abstracts of MAA Seaway Section Meeting, Spring, 2013.

9. W. S. Andrews, *Magic Squares and Cubes* (Dover, New York, 1960).

10. C. Pickover, *The Zen of Magic Squares, Circles and Stars*, Princeton University Press. 2002.

11. H. Behforooz, On the Constructing 4 by 4 Magic Squares with Pre-Assigned Magic Sum, *Mathematical Spectrum*, Vol. 40 (3) 2007/2008.

12. H. Selin (Editor), Yang Hui Magic Squares in Chinese Mathematics, *Encyclopedia of the History of Science, Technology and Medicine in Nonwestern Cultures*, Springer, New York, 1997.

13. Razi Al Din Tabarsi, *Makarim Al-Akhlaq,* (an Old Iranian Book, 12th century).

14. A. Majlesi, *Helyatol Mottaghin*, (an Old Iranian Book, 16th century).

15. N. Namakshi, et al. Mystical Magic Squares, Mathematics Teacher in the Middle School, Vol. 20 (6) 372-377, February 2015

16. E. Pegg Jr., Melencolia Magic, MAA FOCUS, Vol. 34 (6) page 3 and 31, 2015

ALPHAMETICS

edited by Charles Ashbacher

All of the alphametics in this section were created by Charles Ashbacher

1. The following alphametic appeared on the front cover of the book **Alphametics As Expressed in Recreational Mathematics Magazine,** edited by Charles Ashbacher.

```
        RMM
       GONE
        NOT
       ─────
       LOST
```

2. Complex negotiations are taking place with war threatened if Iran develops a nuclear weapon

```
       IRAN
      NUKES      where the goal is to minimize NUKES
        RIP
      ─────
      PEACE
```

3. In 2014, Libya began erupting into a full-blown civil war

```
       2014
      LIBYA
      WARIS
      ─────
      CIVIL
```

The following three alphametics will be appearing in my upcoming book containing alphametics based on episodes from the Star Trek© original series.

4. Tribute to one of the best episodes, number 33, "Mirror, Mirror"

```
          33
        STAR
        TREK      where TREK is the greatest!
      MIRROR
      MIRROR
      ──────
      EMPIRE
```

5. Tribute to episode 1, "The Man Trap."

```
    1
SALT
   IS        where TRAP is the greatest
AMAN
  ----
TRAP
```

6. Tribute to episode 62, "Day of the Dove"

```
   62
DAY
   OF     where the peace makes it the greatest DAY
THE
  ----
DOVE
```

7. Doubly true Basque 1 + 3 + 3 = 7

```
   BAT
   HIRU      where HIRU is also evenly divisible by 3
   HIRU
  ----
   ZAZPI
```

PROBLEMS AND CONJECTURES THAT WOULD HAVE APPEARED IN "JOURNAL OF RECREATIONAL MATHEMATICS" 38(4)

Edited by: Lamarr Widmer

Problems for which no solution is known to the contributor or editor are identified by asterisks (*). Those for which the only known solution requires use of a computer are identified by up-arrows (↑).

1. Pythagorean Dissection by Brian Barwell, Hampton, Middlesex, UK

Find a four-piece dissection of a 7×7 square and a 24×24 square such that the pieces can be re-assembled to form a 25×25 square. Pieces may be turned over but the cuts may only be made along boundaries of the unit squares which make up the larger squares.

2. Pentomino Rectangle with Holes by Brian Barwell, Hampton, Middlesex, UK

Figure 1 shows the eighteen one-sided pentominoes and Figure 2 shows how they can be arranged to form an 8×15 rectangle with a central 6×5 hole. Use these eighteen pieces to construct a rectangle with a central rectangular hole with area greater than 30. The pentominoes may be rotated but not turned over.

3. Wall Scraper by Hubert Hagadorn, Menlo Park, CA

This problem was published in "Topics in Recreational Mathematics 1/2015" but the figure was unfortunately not included. Therefore, it reappears here.

A semicircle of diameter 2 is able to move along a path of unit width having a sharp right angle turn, sliding, rotating and then sliding again as shown in Figure 3. What is the shape of largest area that is able to travel along this path and negotiate the turn?

4. Point of Concurrency in a Square by Subramanyam Durbha, Norristown, PA

Let *ABCD* be a square. Let *E* be the midpoint of *BC*, *F* be the midpoint of *CD* and *G* the midpoint of *BE*. Prove that the lines *AE, BF* and *DG* are concurrent.

Figure 1

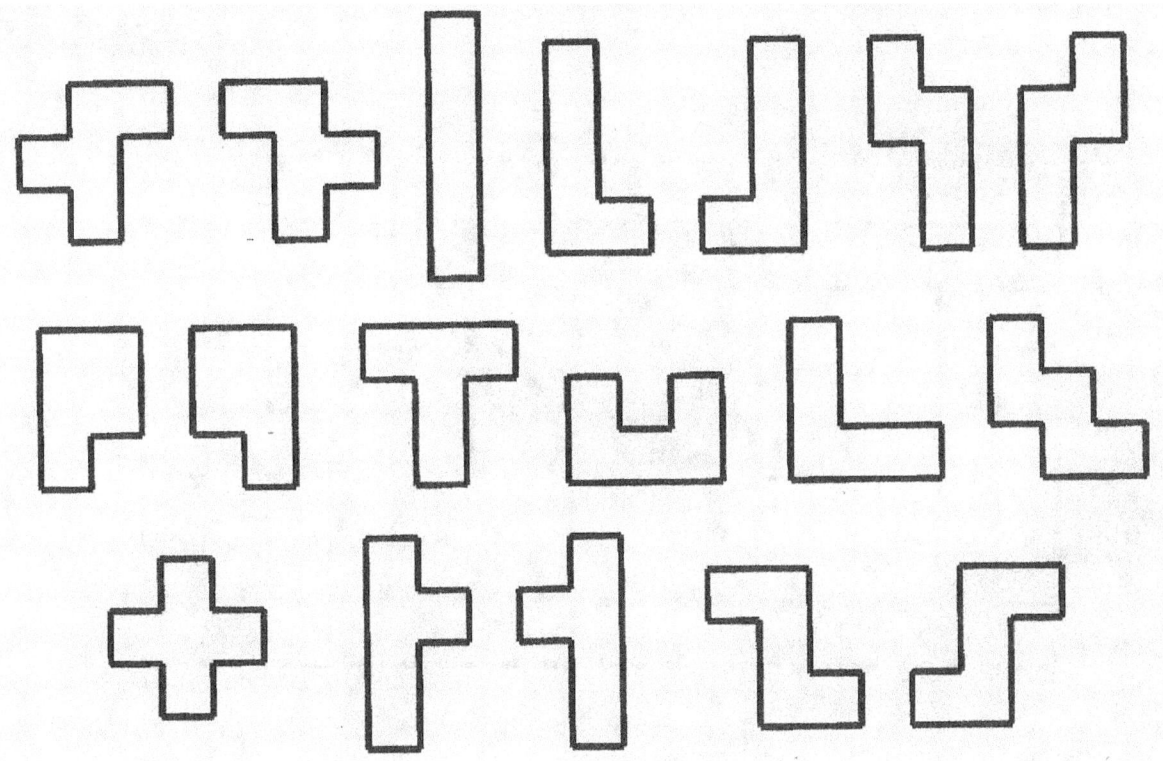

Figure 2

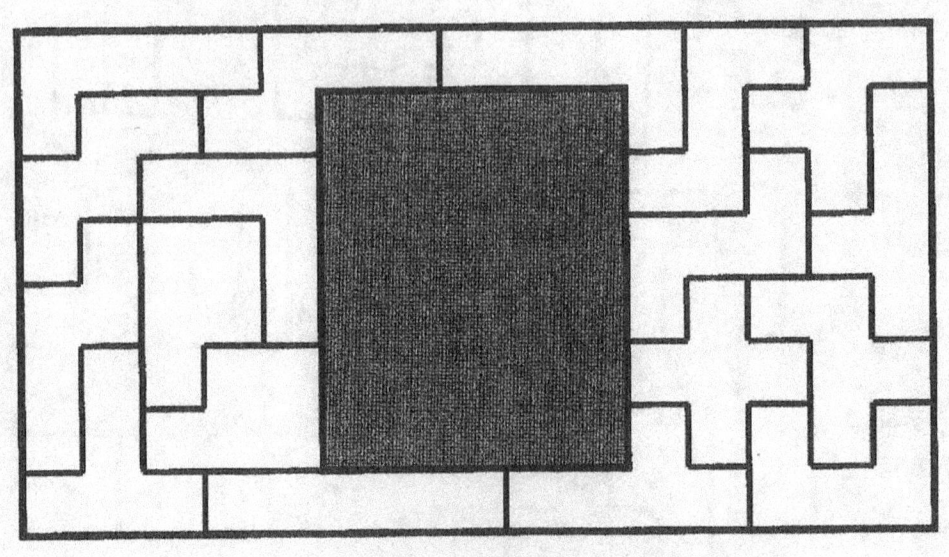

Figure 3

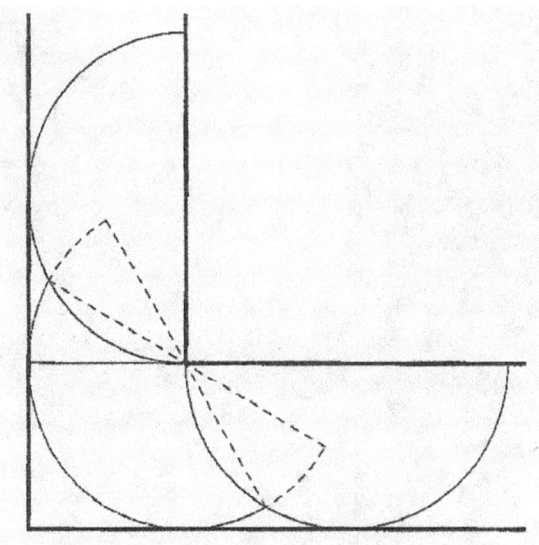

UNFINISHED BUSINESS FROM "JOURNAL OF RECREATIONAL MATHEMATICS," NUMBER 1

by Charles Ashbacher

When I was completing the proofing for the re-publication of the 10-year cumulative index to **Journal of Recreational Mathematics**, I spotted two entries that were listing of problems that ended with "no solution published." My natural reaction was to search for these entries and try to determine why no solution had ever appeared.

The first entry was "Alphametic 411, page 227, Vol. 8(3), no solution published."

I found the problem and the following is a verbatim rendition.

411. A Sad Story

Dad, wife Dal, son Jim, and a daughter married to the church, were on the lam from sin, when Dad tripped and hurt his shin.

In the solution required, both NUN and SIN are prime, but the real challenge is this: In what part of the world did this incident occur?

```
 SAD
 DAD
 DAL
 JIM
 NUN
 LAM
 SIN
-----
SHIN
```

I worked through the problem and when I started finding multiple solutions wrote a quick Java program to count them. The end result was that there were 20 solutions satisfying the conditions, leading me to believe that the reason no solution was published was due to the large number of solutions. Given the standard practice with alphametics, it also seems likely that one additional constraint was missing.

The first solution is

```
  381
  181
  189
  406
  757
  986
  307
 ─────
 3207
```

The second entry corrresponds to a set of problems that appeared in the short paper "Find the Next Term" by N. J. A. Sloane that appeared on page 146 of **Journal of Recreational Mathematics**, Volume 7, Number 2. The problems are repeated as they appeared in that issue and the solutions appear immediately after.

FIND THE NEXT TERM

N. J. A. Sloane

Bell Telephone Laboratories

Murray Hill, New Jersey

Here are some interesting puzzles. Find the next two terms in the following sequences, and the rules for generating them. The first few are easy, so try them on yourself, your friends, your parents and your children. Try the rest on your in-laws and enemies. None require specialized knowledge. Solutions are available on request. These sequences are taken from Reference 1 and were gathered from **Eureka**, the **American Mathematical Monthly**, and other sources.

(1)　1, 4, 9, 16, 25, 36, 49, ...

(2)　1, 3, 6, 10, 15, 21, 28, . . .

(3)　1, 1, 2, 3, 5, 8, 13, 21, 34, ...

(4) 1, 2, 5, 12, 29, 70, 169, 408, . . .

(5) 1, 1, 3, 1, 5, 3, 7, 1, 9, 5, 11, 3, 13, 7, 15, 1, …

(6) 1, 21, 21000, 101, 121, 1101, 1121, 21121, …

(7) 1, 3, 4, 7, 11, 18, 29, 47, 76, 123, . . .

(8) 1, 2, 3, 7, 43, 1807, 3263443, . . .

(9) 1, 4, 11, 20, 31, 44, 61, 100, 121, 144, 171, …

(10) 1, 1, 1, 2, 1, 2, 1, 3, 2, 2, 1, 4, 1, 2, 2, 5, …

(11) 1, 2, 2, 1, 1, 2, 1, 2, 2, 1, 2, 2, 1, 1, 2, 1, 1, 2, 2, 1, 22, 1, 1, 2, 1, 2, 2, 1, 1, 2, 1, 1, 2,

1, 2, 2, 1, 2, 2, 1, 1, …

(12) 4, 6, 7, 9, 10, 11, 12, 14, 15, 16, 17, 18, 19, 20, 22, 23, 24, …

(Find the next 15 terms)

(13) 1, 2, 3, 4, 9, 27, 512, 134217728, …

(14) 1, 1, 2, 1, 2, 2, 3, 1, 2, 2, 3, 2, 3, 3, 4, 1, …

(15) 1, 2, 3, 4, 5, 6, 6, 6, 10, 11, 12, 13, 14, 15, 8, 17, 12, 19, 20, 21, 22, 23, 28, 10, …

(16) 1, 8, 11, 69, 88, 96, 101, 111, 181, 609, 619, …

(17) 1, 2, 4, 10, 26, 76, 232, 764, 2620, 9496, …

(18) 1, 2, 4, 8, 1, 6, 3, 2, 6, 4, 1, 2, 8, 2, …

(19) 1, 3, 5, 8, 12, 18, 24, 30, 36, 42, 52, 60, 68, 78, …

(20) 0, 1, 1, 2, 4, 7, 13, 24, 44, 81, 149, 274, …

(21) 1, 2, 2, 3, 2, 4, 2, 4, 3, 4, 2, 6, 2, 4, 4, 5, …

(22) 4, 6, 9, 10, 14, 15, 21, 22, 25, 26, …

(23) 1, 5, 19, 65, 211, 665, 2059, 6305, 19171, 58025, 175099, …

Reference

1. N. J. A. Sloane, *A Handbook of Integer Sequences*, Academic Press, New York, December, 1973, 220 pages, $10.00, ISBN 0-12-648550-X.

Solutions:

The definitive source for information on integer sequences is The On-Line Encyclopedia of Integer Sequences® (OEIS®), (http://oeis.org/), founded by N. J. A. Sloane. The opening screen has a textbox with the caption

Enter a sequence, word, or sequence number:

Where the user can simply type in the sequence. This page is the source of all the nonobvious answers that are given here.

(1) This is clearly the sequence of the squares of the positive integers.

(2) This is clearly the sequence of integers where the differences between successive numbers are 2, 3, 4, 5, …

(3) This is clearly the set of Fibonacci numbers $F(1) = 1$, $F(2) = 1$, $F(n+1) = F(n) + F(n-1)$

(4) Pell numbers: $a(0) = 0$, $a(1) = 1$; for $n > 1$, $a(n) = 2*a(n-1) + a(n-2)$ (From OEIS)

(5) Remove 2's from n; or largest odd divisor of n; or odd part of n. (From OEIS)

(6) Smallest natural number requiring n words in English (as spoken in England). (From OEIS)

(7) Lucas numbers (beginning at 2): $L(n) = L(n-1) + L(n-2)$. (From IEIS)

(8) A variant of Sylvester's sequence: $a(0)=1$ and for $n>0$, $a(n) = (a(0)*a(1)*...*a(n-1)) + 1$. (From OEIS)

(9) Squares written in base 8.

(10) The multiplicative partition function: number of ways of factoring n with all factors greater than 1 ($a(1)=1$ by convention).

(11) Kolakoski sequence: a(n) is length of n-th run; a(1) = 1; sequence consists just of 1's and 2's (From OEIS)

(12) Non-Fibonacci numbers. Next fifteen terms: 25, 26, 27, 28, 29, 30, 31, 32, 33, 35, 36, 37, 38, 39, 40 (From OEIS)

(13) An exponential function on partitions (next term is 2^512). (From OEIS)

(14) 1's-counting sequence: number of 1's in binary expansion of n (or the binary weight of n). (From OEIS)

(15) Mosaic numbers or multiplicative projection of n. (From OEIS)

(16) Strobogrammatic numbers: the same upside down. (From OEIS)

(17) Number of self-inverse permutations on n letters, also known as involutions; number of Young tableaux with n cells. (From OEIS)

(18) Digits of powers of 2. (From OEIS)

(19) a(1) = 1, for n>=2: a(n) = sum of two consecutive noncomposite numbers (From OEIS)

(20) Tribonacci numbers: a(n) = a(n-1) + a(n-2) + a(n-3) with a(0)=a(1)=0, a(2)=1. (From OEIS)

(21) d(n) (also called tau(n) or sigma_0(n)), the number of divisors of n. (From OEIS)

(22) Semiprimes (or biprimes): products of two primes. (From OEIS)

(23) 3^n - 2^n. (From OEIS)

It has been a goal of mine for some time to have solutions created for all problems that have appeared in **Journal of Recreational Mathematics**. As I go through back issues and encounter problems, I will be trying to develop solutions. Therefore, this article is hopefully the first of what will become a regular feature until there is no more reason to continue it.

BOOK REVIEWS

Edited by: Charles Ashbacher

Charles Ashbacher Technologies

5530 Kacena Ave

Marion, IA 52302

E-mail: cashbacher@yahoo.com

When Life Is Linear: From Computer Graphics to Bracketology, by Tim Chartier, The Mathematical Association of America, Washington, D. C., 2015. 140 pp., $50.00 (paper). ISBN 978-0883856499.

Linear systems are generally easy to understand, so they are found in nearly all basic algebra texts. Given the complexity of life it is surprising, often even to mathematicians, that linear equations in the form of matrices can be used to accurately describe many natural phenomena. Chartier describes some of the most common uses for matrices, from computer graphics, to encryption, to complex algorithms used by Google to rank pages, and finally to filling out a winning NCAA March Madness bracket.

The level of difficulty is not high; a college course in linear algebra is not even necessary for comprehension. The only background needed is an understanding of more complex linear systems, some trigonometry, and the basic operations on matrices. Segments could be pulled out and used as supplements for high school courses, college courses in finite mathematics, and to answer the standard question, "What will we ever use math for?"

Readers will best relate to the sections on computer graphics and March Madness. Nearly all people are familiar with the power of computer graphics from watching movies, and the number of NCAA basketball tournament office pools is enormous. Even someone that does not know matrix algebra should be able to understand the section on the relative ranking of teams.

Math can be fun and it can be useful. A small percentage of people consider it both. With the existence of this book, that percentage will rise.

Charles Ashbacher

Cows in the Maze, by Ian Stewart, New York, Oxford University Press Inc., 2010, 306 pp. $17.95 (paper), ISBN 978-0-19-956207-7

The material in this book originally appeared in the author's "Mathematical Recreations" column in *Scientific American* and *Pour la Science*. Each of the chapters is an updated version of one such column. The updating consists mostly of feedback from readers of the columns and Stewart's response in some cases. The chapters end with a list of websites with further material on their topics.

This book's title refers to chapter 13, where Stewart presents an ingenious and difficult maze invented by Robert Abbott, who published it in his book, *Supermazes*. The maze is in the form of a flowchart through which two markers move. The player can choose which marker to move at each turn, and the goal is to have one marker reach the designated final position. In most boxes, a logical "if" statement directs the player to one of two exits based on the current state of the game. Success depends in no small part on careful attention to the logic of these statements. A glance at the flowchart easily shows there is only one box through which the final box can be entered. The successful solver must realize at some point that the only way for one marker to reach the final box is for both to simultaneously occupy this penultimate box. Like a mathematical proof, this puzzle can be solved by reasoning forward from the initial state to the desired final state, backward from the final state or, most likely, some combination of both.

Other chapters in this collection deal with conditional probability, topology, tessellations, magic squares, knight's tours, and the lore of dice. Stewart is certainly not the first to write about these topics, but in every case he brings something new and his presentation is engaging. He also includes a number of less familiar topics such as a study of real knots. This is not knot theory in the usual purely topological sense; it takes into account the actual thickness of the string in a real-world knot and the friction of the string against itself. Chapter 16 tells of a glassblower who gained new insights by constructing glass models of Klein bottles and other surfaces. I especially like the pictures which accompany the text of this chapter. One such photo is available in color at http://www.sciencemuseum.org.uk/objects/mathematics/1996-545.aspx . The text that accompanies it prompts a question: how can one Klein bottle be "inside" another (since a Klein bottle does not have an inside)?

The material in this book is diverse, accessible, clear, and appealing. I enjoyed it and certainly will be recommending it to students in the future

Lamarr Widmer

Crimes and Mathdemeanors, by Leith Hathout, Wellesley, Massachusetts: A. K. Peters, 2007. 196 pp., $14.95 (paperbound), ISBN 1-56881-260-4

This work consists of fourteen short chapters with a consistent organization. Each one begins with a short fictional account of a puzzling situation, in most cases involving some sort of criminal activity. The account is interrupted at a critical juncture and the reader is invited to solve the mystery. Some logical or mathematical analysis is needed. The reader may choose to work with the clues concealed in the fictional account, or they may proceed to read the author's analysis, which highlights the relevant information and clearly formulates the mathematical problem to be solved. The reader may then choose how long to work on the problem before moving on to Hathout's solution. These solutions are clearly presented, generally at an appropriate level for a high school or college student.

The protagonist in these fictional mysteries is Ravi, a 14-year with strong observational skills and considerable mathematical ability. He is in many ways a typical teenager with a love of basketball and a taste for donuts. Ravi's father is a district attorney in Cook County, Illinois. The chief of the Criminal Investigations Unit of the Chicago Police Department is the father of one of Ravi's tenth grade classmates. Both adults seek Ravi's help with particularly puzzling crimes. The very first crime in this book is a murder case in which Ravi deduces the identity of the killer based on the evidence, including the statements of the likely suspects. There is absolutely no gratuitous violence in this or any other case in this book. Other cases involve fraud, theft, and an incident in which Ravi rescues two schoolmates who are unjustly accused of cheating by a teacher. Ravi shows proper respect for the teacher while unavoidably showing himself to be intellectually superior.

This book is remarkable for a number of reasons. The vignettes are imaginative and amusing. The solutions employ an impressive variety of mathematical topics including algebra, combinatorics, geometry, analytic geometry, calculus, and differential equations. One solution uses Blichfeldt's Lemma, which was new to me. The author provides an ingenious, intuitively attractive proof of this lemma before applying it to solve a murder. Finally, and I purposely saved this for last, Leith Hathout himself is a high school math whiz with a record of success in mathematics competitions. My regular readers know my appreciation for the work of the many talented popularizers who contribute to mathematical literature. I welcome the appearance of this impressive and promising newcomer to the genre.

Lamarr Widmer

Loving + Hating Mathematics, by Reuben Hersh and Vera John-Steiner, Princeton NJ, Princeton University Press, 2011, 416 pp. $29.95 (hardcover), ISBN 978-0-691-142470

As the book jacket says, this work "is about the hidden human, emotional, and social forces that shape mathematics and affect the experiences of students and mathematicians." This is a book full of stories – stories about people, with scant mention of mathematical problems or theorems. It covers stages in the life of a mathematician, partnerships of mathematicians, mathematical communities, questions of gender and age in the mathematical profession, and, in its final two chapters, issues related to mathematics education.

Chapter 1 is about the early awakening of mathematical interest in budding talents such as Gauss, Stan Ulam, Andrew Wiles, and Sonia Kovalevskaya. Representative of the material here is a childhood reminiscence of Anneli Lax, who said mathematics was "the perfect sort of escape; I didn't have to look up anything; I didn't have to consult libraries or books. I could just sit there and figure things out." This chapter also examines the phenomenon of childhood prodigies and the influence of teachers on young mathematical talent.

The wide-ranging Chapter 2, titled Mathematical Culture, includes a section on mathematical beauty as well as a fascinating account of the tension between Grigori Perelman and Shing-Tung Yau over Perelman's proof of the Poincaré conjecture. This chapter ends with a consideration of issues related to academic tenure, focusing in particular on the case of Jenny Harrison at Berkeley. Chapter 3 considers the solace that some find in mathematics.

The provocative Chapter 4 considers eccentricities, obsessions, and mental illness. On page 107 the authors say, "For some mathematicians, it seems, mathematics can be a destructive addiction. In some cases, there was a specifically mathematical style or flavor to their delusions." Here the authors mention John Nash, Kurt Gödel, and André Bloch. We are struck by the remarkably logical structure of the "axioms" and conclusion of the manifesto of Ted Kaczynski (the Unabomber), holder of a Ph.D. in mathematics from the University of Michigan. The authors devote twenty pages to the enigmatic life of Alexandre Grothendieck, former mathematics superstar who is now totally alienated from the mathematical community, living as a hermit in a remote French village.

Chapter 5 is more uplifting with accounts of mathematical friendships and partnerships, including the unlikely relationship of Karl Weierstrass and Sonia Kovalevskaya. Other partnerships include Hardy/Littlewood, Hardy/Ramanujan, and Gödel/Einstein as well the marital and professional relationship of Julia and Raphael Robinson. Chapter 6 considers a number of mathematical communities such as the Courant Institute and the Jewish People's University, the latter existing briefly and ending tragically. Chapter 7 examines issues of gender and age, notably G. H. Hardy's "mathematics is a young man's game" claim, which the authors conclude is misleading, even harmful.

The final two chapters consider mathematics education, including questions of curriculum reform and dealing with "reluctant learners" (where the hatred in the title comes in), both pivotal issues for the future of our discipline. I was captivated by every page of this very well-written book. I highly recommend it to all those who are interested in the cultural side of mathematics.

Lamarr Widmer

A Mathematical Medley: Fifty Easy Pieces on Mathematics, by George C. Szipiro, Providence, RI, American Mathematical Society, 2010, 236 pp. $35.00 (paper), ISBN 978-0-8218-4928-6

This popularization contains translations of 41 stories from two German publications, plus nine new stories. This is easy reading; almost no math background is required and these short essays do not present theorems or challenging calculations. The fifty chapters are independent, allowing the reader to sample in any order. They are presented in nine sections, grouped by subject matter. One section, titled Personalities, includes seven biographical sketches of some fascinating but little know mathematicians, including Bella Abramova Subbotovskaya and

Stephen Smale. Other sections are titled Training the brain, Choosing and dividing, and Interdisciplinary matters.

The section titled Games, gifts and other diversions may be of greatest interest to most readers. The eight essays in this section include ones covering mathematical issues related to Rubik's cube, Sudoku, and Tic-tac-toe. The essay Liars and Half-liars considers a guessing game (like twenty questions) in which the holder of the secret is allowed a predetermined number of false answers.

The writing style is clear and entertaining, evidently a reflection of the original German versions. The great variety of topics from mathematics and statistics is a strong point, and even the more familiar topics are freshened with a new details or viewpoints.

Lamarr Widmer

The Numbers Game, by Michael Blastland and Andrew Dilnot, New York, Gotham Books, 2009, 210 pp. $22.00 (hardcover), ISBN 978-1-592-40423-0

This book is a worthy new contribution to the literature on *innumeracy*, the topic of incompetence, misunderstanding, and misuse of numbers. It gains its credibility and relevance by drawing its material from recent public discourse. Most of the examples found here belong to the realm of statistics and are taken from the popular media and political discussion.

Particularly interesting to me is the authors' discussion of numbers with many digits in chapter two. I purposely have avoided saying "large numbers" in the previous sentence because that is precisely their point: a number is not necessarily *big* just because it ends in a long string of zeros. They discuss an example in which people were asked to estimate the total UK government expenditure on health service in the year 2005 and choose their answer from a number of proposed possibilities. Some chose £7 million, which, as the authors point out, is the price of a large home in certain exclusive parts of London. Those who opted for £70 million did not realize that this comes to £1.20 per resident. This illustrates the authors' main recommendation for understanding the significance and plausibility of such numbers: divide them by the total population. Presumably in this example, that should lead one to the correct number, which is £70 billion.

Other chapters of this work include material relating to sampling, recognizing outliers, assessing their significance, false conclusions of causation, and distinguishing significant trends from random variations. In all cases, the presentation includes relevant, interesting, and current examples whose sources are cited. The authors somehow manage to maintain an admirably balanced tone while working with such politically charged topics as nationalized health care.

Blastland and Dilnot are creator and former host of a BBC radio show called *More or Less*. Much of the material in this book is from episodes of that show. A version of this book was previously published in the UK with the title *The Tiger That Isn't*. The jacket states that this new edition is "comprehensively updated and adapted specifically for American readers."

Lamarr Widmer

Math Goes to the Movies, by Burkard Polster and Marty Ross, Baltimore, MD, The Johns Hopkins University Press, 2012, 286 pp. $35.00 (paper), ISBN 978-1-4214-0484-4

These authors are collectors and cataloguers. Their process of collecting information on mathematical episodes in popular cinema is ongoing and their website (www.qedcat.com) lists more than 800 math movies they have seen. The site provides summaries for every one as well as information on where they may be purchased and links to clips. In their Preface, the authors note that some of these movies are not particularly entertaining. They have sifted through all of them and this book presents some of their best and most significant material.

Part 1 of this work consists of twelve chapters, each of which examines one particular movie. The first three cover *Good Will Hunting, A Beautiful Mind,* and *Stand and Deliver*. Polster and Ross are not neutral; they tell us what they like and whether the math in these movies is correct or faulty. Chapter 5 covers *Donald in Mathmagic Land,* which they criticize for its inclusion of a large amount of nonsense related to the golden ratio. Being familiar with this work, I would have been shocked if they had done otherwise. They also point out bloopers in several of the movies they considered. I suspect the general audience is oblivious to many of these, while the mathematically inclined viewer is amused by their absurdity. Some of the best of these movies employed mathematicians as consultants in order to ensure mathematical authenticity. In chapter 2, Polster and Ross highlight the work of Professor David Bayer, consultant for *A Beautiful Mind,* who provides insight from his unique perspective on the production of this highly acclaimed work.

Part 2 consists of similar material, this time organized by topic. For example, Chapter 14 covers appearances of the Pythagorean Theorem and Fermat's Last Theorem in a number of movies. Chapter 18 focuses particularly on bloopers, although they are mentioned elsewhere as well. Part 3 consists of almost fifty pages of lists, useful for those seeking particular types of mathematical references in movies. For example, section 21.8 lists thirteen movies containing references to pi. Finally, the usefulness of this work is enhanced by a Movie Index listing every movie and the pages on which it is mentioned.

Regular readers of my reviews are likely aware of my interest in popularization of mathematics and the image of mathematics in the popular media. The topic of mathematics in movies and television has been popular with my students, and I have learned from them and from my own viewing. However, I would be hard pressed to name more than twenty-five or thirty movies with any reference to mathematical ideas. I had no idea there could be more than eight hundred! I believe this book and the corresponding website are the most extensive source on the topic of mathematics in movies. They are a welcome addition to earlier sources, many of which can be reached by links from Polster's and Ross's website.

Lamarr Widmer

NAKED WORDS: The Effective 157-Word Email, by Gisela Hausmann, $2.99 on Amazon Kindle, http://www.amazon.com/NAKED-WORDS-Effective-157-Word-Email-ebook/dp/B00S6YNM7A/ref=sr_1_1?s=books&ie=UTF8&qid=1426864016&sr=1-1&keywords=Gisela+Hausmann

Not too long ago, I sent an email to the property manager at an apartment complex inquiring about possible openings. Her response was prompt but instantly sent the wrong impression. Her email was written in pink and purple Comic Sans. The response was unnecessarily abrupt, contained no greeting or closing, and neglected to address all of my questions. I hadn't even met the property manager yet, but I already had my first impression of her: unprofessional.

In the digital age, it's easy to take the simple act of writing an email for granted. Gisela Hausmann's short and sweet book, *Naked Words*, compiles helpful tips, dos and don'ts, and examples of good and bad emails, all in one place for easy reference. Although the organization of her book feels a little haphazard and unnatural in places, Hausmann makes a lot of use out of just a few pages. She clues the reader in to recent statistics about emails and provides clear instructions on how to write an email that sets one apart from the competition.

What I enjoyed most about the book was the section filled with examples of truly horrendous emails. The emails themselves are a combination of hilarious and painful-to-read disasters, but Hausmann's playful, cutting commentary is what really made me laugh. I think I could read a whole book of the author's sassy reflections on unprofessional emails.

Anyone who works in a business or academic setting should refer to *Naked Words*. It's a quick and helpful read. Even if you consider yourself a master of writing clear, concise, and unique emails, I would greatly recommend taking a peek at this book for a quick refresher. A little tweak to your emails can be the difference between rising above the competition and fading into obscurity.

Jennifer Corrigan

SOLUTIONS TO PROBLEMS AND CONJECTURES FROM JOURNAL OF RECREATIONAL MATHEMATICS 37(4)

Edited by: Lamarr Widmer

2860. Stanford Steps by Don Knuth, Stanford, California (*JRM* 37:4, p. 352)

A certain university don climbs eight flights of stairs every day when he goes to his office, where individual flights consist of 4, 9, 5, 11, 11, 12, 11, 12 steps, respectively. One day he notices that he could climb the stairs in groups 1+2+1, 2+1+2+1+2+1, 2+3, 2+1+2+3+2+1, 2+1+2+1+2+1+2, 3+2+1+2+1+2+1, 2+1+2+1+2+1+2, 1+2+1+2+1+2+1+2, always mounting either 1, 2 or 3 steps at a time and alternating between even and odd. In how many different ways can the don climb to his office with such a procedure?

Solution by Daniele Degiorgi

Let $f_{i,j}(n)$ be the number of ways to climb a flight of n steps where $i \in \{0,1\}$ is the parity of the first group and j is the parity of the last group. We have
$f_{1,1}(1) = f_{0,0}(2) = f_{1,1}(3) = f_{0,1}(3) = f_{1,0}(3) = 1$ while all other values for $n < 4$ are 0. Then we have the recursion relations $f_{0,k}(n) = f_{1,k}(n-2)$ and $f_{1,k}(n) = f_{0,k}(n-1) + f_{0,k}(n-3)$. With a few calculations we get the following.

n	$f_{0,0}(n)$	$f_{0,1}(n)$	$f_{1,0}(n)$	$f_{1,1}(n)$
1	0	0	0	1
2	1	0	0	0
3	0	1	1	1
4	0	0	0	1
5	1	1	1	0
6	0	1	1	2
7	1	0	0	1
8	1	2	2	1
9	0	1	1	3
10	2	1	1	1
11	1	3	3	3
12	1	1	1	4

Our answer is thus

$$\sum_{i_k \in \{0,1\}} f_{i_1,i_2}(4) f_{1-i_2,i_3}(9) f_{1-i_3,i_4}(5) f_{1-i_4,i_5}(11) f_{1-i_5,i_6}(11) f_{1-i_6,i_7}(12) f_{1-i_7,i_8}(11) f_{1-i_8,i_9}(12) = 2130$$

We note that most of the 512 terms in this sum are 0.

2861. Concealed Choices by Frank Rubin, Wappingers Falls, New York (*JRM* 37:4, p. 353)

Four mathematicians decided to play a game. Each one picked a positive integer. Then they took turns finding relationships among their numbers. The first one announced, "There is a unique way in which powers of your numbers add to a 40-digit power of my number." The second one revealed, "There is a unique way in which powers of your numbers add to a 41-digit power of my number." The third one stated, "There is a unique way in which powers of your numbers add to a 42-digit power of my number." The last one declared, "There is a unique way in which powers of your numbers add to a 43-digit power of my number." What numbers had they chosen?

Solution by Hubert Hagadorn

If the numbers are $(a,b,c,d) = (2,4,8,16)$, then there are three solutions. For the first the choosers speak in the order (a,b,c,d) and their statements correspond to the equalities
$2^{130} = 4^{64} + 8^{43} + 16^{32}$, $4^{67} = 2^{133} + 8^{44} + 16^{33}$, $8^{46} = 2^{137} + 4^{68} + 16^{34}$ and
$16^{35} = 2^{139} + 4^{69} + 8^{46}$. For the second solution, they speak in the order (a,d,c,b) and their statements correspond to the equalities
$2^{130} = 4^{64} + 8^{43} + 16^{32}$, $16^{34} = 2^{134} + 4^{67} + 8^{45}$, $8^{46} = 2^{137} + 4^{68} + 16^{34}$ and
$4^{71} = 2^{140} + 8^{47} + 16^{35}$. For the third solutions, they speak in the order (b,d,c,a) and the equalities are $4^{65} = 2^{128} + 8^{43} + 16^{32}$, $16^{34} = 2^{134} + 4^{67} + 8^{45}$, $8^{46} = 2^{137} + 4^{68} + 16^{34}$ and
$2^{142} = 4^{70} + 8^{47} + 16^{35}$.

If the numbers are $(a,b,c,d) = (3,3,3,27)$, then there are six solutions. In each of them, a, b and c speak in any order, followed by d. The equalities, in any case, are $3^{82} = 3^{81} + 3^{81} + 27^{27}$,
$3^{85} = 3^{84} + 3^{84} + 27^{28}$, $3^{88} = 3^{87} + 3^{87} + 27^{29}$ and $27^{30} = 3^{89} + 3^{89} + 3^{89}$.

2862. XXX-Rated by Lamarr Widmer, Mechanicsburg, PA (*JRM* 37:4, p. 353)

Use nine of the twelve pentominoes (Figure 1) to construct the triple scale X shown in Figure 2.

Figure 1

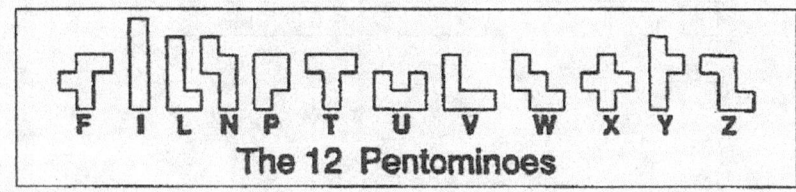

The 12 Pentominoes

Figure 2

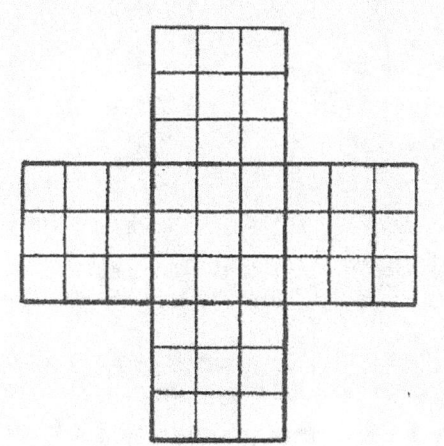

Solution by Daniele Degiorgi

Eleven solutions are shown in Figure 3.

Figure 3

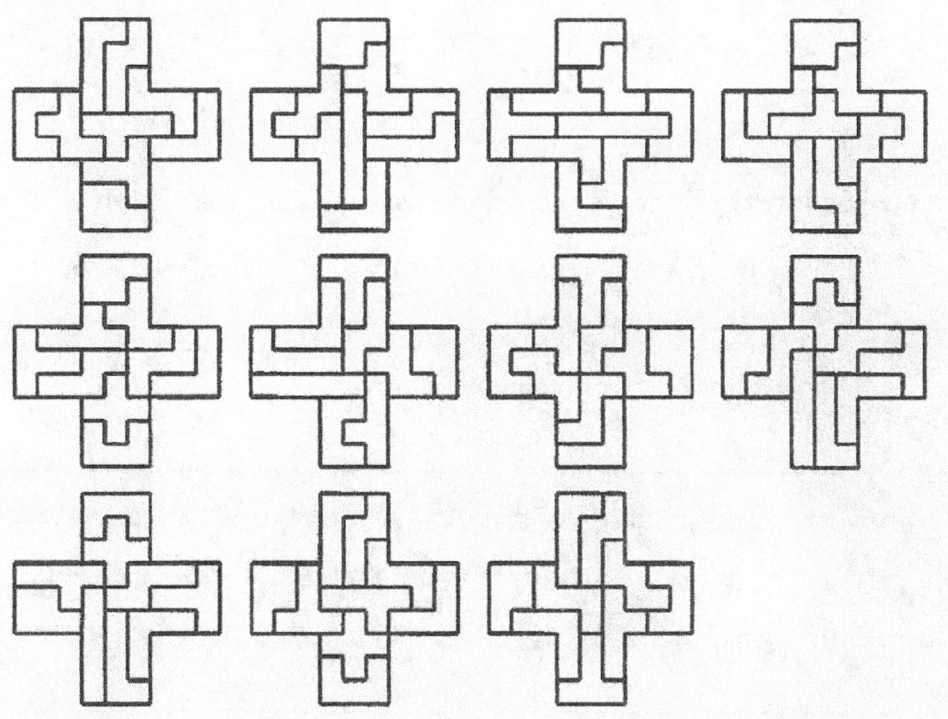

2863. Matrix Square Roots by Phoebe Chua, Mechanicsburg, PA (*JRM* 37:4, p. 354)

Find all solutions of the matrix equation $A^2 = \begin{bmatrix} 2 & 2 \\ 2 & 2 \end{bmatrix}$.

Solution by Michael P. Cohen

$$A^2 = \begin{bmatrix} a & b \\ c & d \end{bmatrix} \times \begin{bmatrix} a & b \\ c & d \end{bmatrix} = \begin{bmatrix} a^2 + bc & ab + bd \\ ac + cd & bc + d^2 \end{bmatrix}.$$

Looking at the main diagonal terms, we have $a^2 + bc = d^2 + bc$ so $d = \pm a$. But if $d = -a$, then the off-diagonal entries are 0, not 2, so $d = a$. Similarly, if $d = a = 0$, then the off diagonal entries are 0, so $a \neq 0$. Equating the off diagonal entries, we have $2ab = 2ac$ so $c = b$. Now we have

$$A^2 = \begin{bmatrix} a^2 + b^2 & 2ab \\ 2ab & a^2 + b^2 \end{bmatrix}.$$

Then $b \neq 0$ and $2 = 2ab$ so $b = 1/a$. Looking at the main diagonal terms, we have $a^2 + 1/a^2 = 2$. Then $0 = a^4 - 2a^2 + 1 = (a^2 - 1)^2$. Therefore $a = \pm 1$ and $b = a$. Finally,

$$A = \begin{bmatrix} 1 & 1 \\ 1 & 1 \end{bmatrix} \text{ or } A = \begin{bmatrix} -1 & -1 \\ -1 & -1 \end{bmatrix}.$$

2864. Hexomino Squares by Dario Uri, Pontecchio Marconi, Italy (*JRM* 37:4, p. 354)

The set of 35 hexominoes is shown in Figure 4. Use 30 distinct hexominoes to build five 6 × 6 squares.

Figure 4

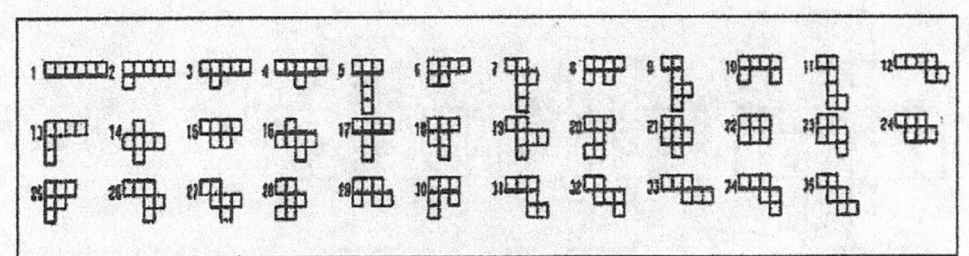

Solution by Daniele Degiorgi

Two solutions are shown in Figure 5.

Figure 5

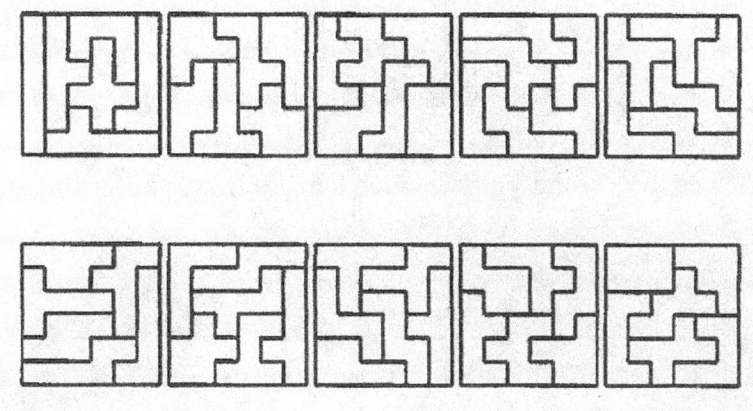

2865. Arresting Resistors by Hubert Hagadorn, Menlo Park, CA (*JRM* 37:4, p. 354)

When n resistors are wired in series as in Figure 6, the total resistance between the two terminals (endpoints) is $R = r_1 + r_2 + \cdots + r_n$. When wired in parallel as in Figure 7, the total resistance is given by

$$\frac{1}{R} = \frac{1}{r_1} + \frac{1}{r_2} + \cdots + \frac{1}{r_n}$$

or equivalently

$$R = \frac{1}{\frac{1}{r_1} + \frac{1}{r_2} + \cdots + \frac{1}{r_n}}.$$

Figure 6

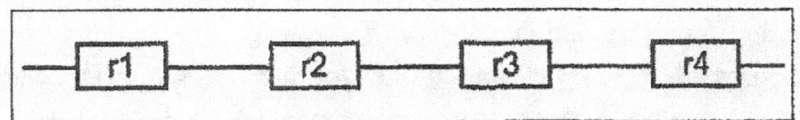

Figure 7

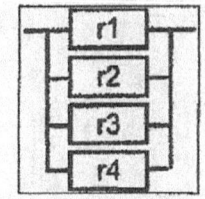

For more complicated circuits both formulas come into play, e.g. for the circuit shown in Figure 8,

$$\frac{1}{R} = \frac{1}{r_1 + r_2} + \frac{1}{r_3}.$$

a. Assume we may select four resistors and specify their resistance values. Then, we design a circuit between two terminals using one or more of these resistors in any possible configuration. How many different resistance values can be achieved between the terminals?
b. Assuming that the resistance values of the four resistors are positive integers, what is the smallest possible sum of their values which allow us to achieve the answer found in part a.?

Figure 8

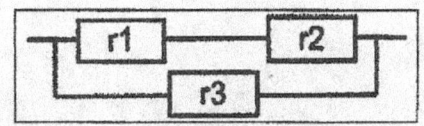

Solution by Daniele Degiorgi and the Proposer (Composite)

a. Using one resistor, we have four values r_1, r_2, r_3, r_4. Using two resistors, we have six possible pairs, each of which can be connected in series and in parallel, for a total of 12 possibilities. With three resistors, we have four possible values when they are wired in series and four more when wired in parallel. For the two mixed cases (Figures 8,9), we have four choices of the three resistors, and then three different ways to wire then, for a total of 12 possibilities. So, in all, there are 32 possibilities for three resistors. There are ten distinct ways (Figure 10) to connect four resistors and they yield 1,1,3,3,4,4,6,6,12,12 for a total of 52. So there are 100 possibly different, resulting resistance values.
b. Since $1 + 2 = 3$, the set $(1,2,3,n)$ cannot give the desired distinct values. Similarly, the set $(1,2,4,n)$, fails for values of n from 5 through 10 (e.g. for $n = 8$, we find that for the

configuration shown in Figure 9, $\frac{1 \cdot 2}{1+2} + 4 = \frac{14}{3} = \frac{4 \cdot 8}{4+8} + 2$). When the values (1,2,5,9) are used, all 100 configurations have a distinct resistance and therefore, our answer is 17.

Figure 9

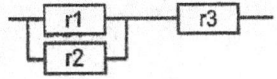

Figure 10

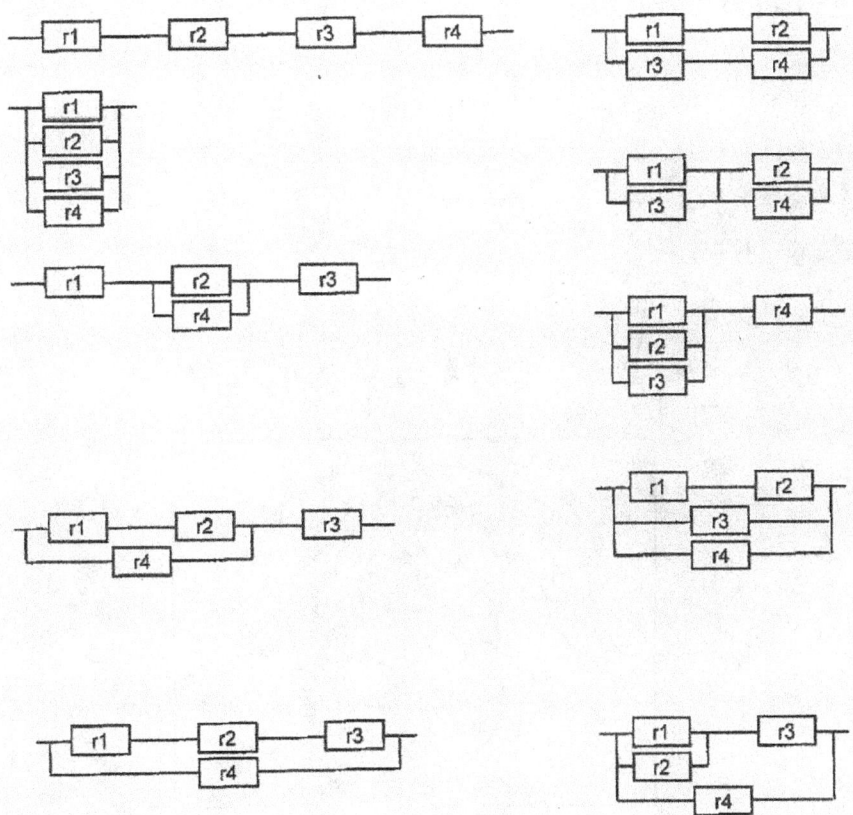

2866. Quadratic Function and Arithmetic Sequence by John Runkle, Lima, Ohio (*JRM* 37:4, p. 355)

For the quadratic function $f(x) = ax^2 + bx + c$, prove that there is a constant C and an arithmetic sequence (s_n) such that for positive integer values n, $f(n)$ is equal to C plus the nth partial sum S_n of the arithmetic sequence.

Solution by Andy Pepperdine

The result follows immediately from

$$\sum_{i=1}^{n} (2ai + (b-a)) + c = (an(n+1) + (b-a)n) + c = an^2 + bn + c$$

2867. Magic Graph by Dario Uri, Pontacchio Marconi, Italy (*JRM* 37:4, p. 355)

Replace the letters A through N at the vertices of the graph shown in Figure 11, with the integers 1 through 14, in such a way that the sums of numbers on the six straight line segments and around the three ellipses are all equal. This means they satisfy

$$A + B + E = A + D + F = A + C + G = E + J + L = F + K + N = L + M + N = A + J + K = B + C + D + H + I = G + H + I + M$$

.

Figure 11

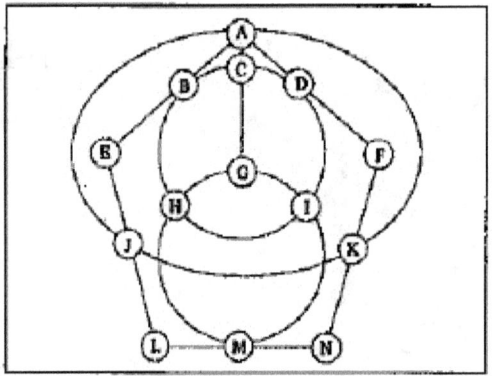

Solution by Henry Ibstedt

$(A, B, C, D, E, F, G, H, I, J, K, L, M, N) = (12,3,4,8,11,6,10,2,9,1,13,14,5,7)$. Another solution is obtained by interchanging 2 and 9.

2868. Drunken Travelling Salesman by Hubert Hagadorn, Menlo Park, CA (*JRM* 37:4, p. 356)

A salesman makes a round trip visit of nine cities spaced on a unit grid as shown in Figure 12. No city is visited more than once and no three cities which lie on a straight line are visited consecutively. What are the minimum and maximum possible lengths of such a round trip?

Solution by Daniele Degiorgi

The minimum and maximum are 12.7148 and 18.0667. Numbering the cities as $\begin{bmatrix} 1 & 2 & 3 \\ 4 & 5 & 6 \\ 7 & 8 & 9 \end{bmatrix}$, one tour of minimum length visits the cities in the order 127635984 and a tour of maximum length is 167294385.

Figure 12

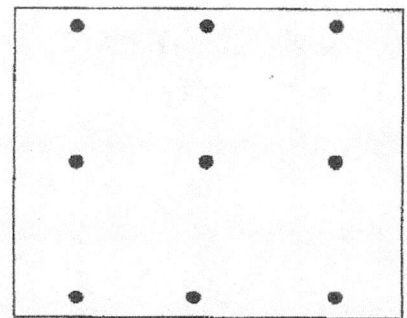

2869. Box of Balls by Frank J. Swetz, Middletown, PA (*JRM* 37:4, p. 356)

In Figure 13, if the radius of each inscribed circle is 1, what are the dimensions of the bounding rectangle? The centers of four of these circles are collinear. This is problem 12 on page 107 of [1], used with permission.)

1. F. Swetz, *Mathematical Expeditions: exploring word problems across the ages*, The Johns Hopkins University Press, Baltimore, 2012

Figure 13

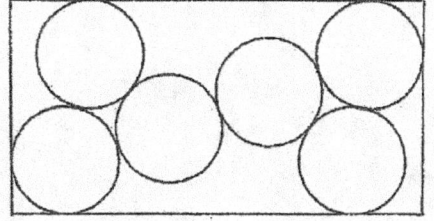

Solution by Daniele Degiorgi

We assume that the lower left corner of the rectangle is at the origin so that the center of the lower left circle is $(1,1)$. We let λ denote the slope of the line through the four concentric centers. So these centers are $(1,1)$, $(1 + 2\cos\lambda, 1 + 2\sin\lambda)$, $(1 + 4\cos\lambda, 1 + 4\sin\lambda)$ and $(1 + 6\cos\lambda, 1 + 6\sin\lambda)$. The center of the upper left circle is $\left(1 + 2\cos\left(\lambda + \frac{\pi}{3}\right), 1 + 2\sin\left(\lambda + \frac{\pi}{3}\right)\right)$ and the center of the lower right circle is $\left(1 + 4\cos\lambda + 2\cos\left(\frac{\pi}{3} - \lambda\right), 1 + 4\sin\lambda - 2\sin\left(\frac{\pi}{3} - \lambda\right)\right)$. Thus the bounding rectangle has width $2 + 6\cos\lambda$ and height $2 + 6\sin\lambda$.

As the centers of the lower left circle and lower right circle have the same y-coordinate, we have $4\sin\lambda - 2\sin\left(\frac{\pi}{3} - \lambda\right) = 0$ which reduces to $5\sin\lambda - \sqrt{3}\cos\lambda = 0$. The solution of this equation is $\sin\lambda = \frac{\sqrt{21}}{14} \approx 0.3273268354$ and $\cos\lambda \frac{5\sqrt{7}}{14} \approx 0.9449111825$. Thus the bounding rectangle has width $2 + \frac{15}{7}\sqrt{7} \approx 7.669467095$ and height $2 + \frac{3}{7}\sqrt{21} \approx 3.963961013$.

SOLUTIONS TO ALPHAMETICS THAT APPEARED IN JOURNAL OF RECREATIONAL MATHEMATICS 37(4)

Edited by Steven Kahan

2849. Words to live by by A. G. Bradbury, North Bay Ontario

```
WASTE + NOT + WANT + NOT = READER
```

The least WASTE is desired.

92470 + 537 + 9257 + 537 = 102801

2850. The Truth About Youth by B. J. Portz, Wauswatosa, WI

```
SOME + ARE + TALL + SOME + ARE = SMALL
```

Solve in base 8 and make SMALL smallest.

1604 + 574 + 3522 + 1604 + 574 = 10522

2851. Some Are Tall, Some Are Small by B. J. Portz, Wauwatosa, WI

```
THE + TRUTH + ABOUT = YOUTH
```

Solve in base 9 and make YOUTH greatest.

207 + 21820 + 34682 = 56820 with 1 & 4 interchangeable.

2852. Pandigital Contrast by Frank Mrazik, Montreal, Canada

```
2(BLACK) = 19(WHITE)
```

2(97356) = 19(10248)

2853. Distinct Summands by Giulio Cesare, Rome, Italy

```
OTTO + NOVE + UNDICI + SEDICI + VENTISEI + VENTOTTO + CENTODUE =
                DUECENTO
```

(8 + 9 + 11 + 16+ 26 + 28 + 102 = 200)

3003 + 9321 + 798545 + 618545 + 21905615 + 21903003 + 41903871 =
    87141903

2854. Repeated Summands by Junya Kanagawa, Japan

```
3666315(ONE) + 30335(ELEVEN) = FOURMILLION
```

3666315(156) + 30335(676365) = 21089477415

2855. Annus Mirabilis by Andrzej Bartz, Fuerth, Germany

$$DCV + DXL + CCV + DLV = MCMV$$

$$E = M \times C^2$$

Solve this pair simultaneously to commemorate Einstein's well-known discovery published in 1905.

$325 + 360 + 225 + 305 = 1215 \quad 4 = 1 * 2^2$

2856. Equal Opportunity – 1 by Andrzej Bartz, Fuerth, Germany

```
SIX + SEVEN + NINETEEN + SEVENTY =

TEN + FIFTEEN + SIXTEEN + SIXTYONE
```

658 + 62327 + 75724227 + 6232741 = 427 + 9594227 + 6584227 + 65841072

2857. Equal Opportunity – 2 by Andrzej Bartz, Fuerth, Germany

```
TWENTYONE + THIRTYONE + THIRTYNINE =

  THIRTEEN + NINETEEN + TWENTYNINE + THIRTY
```

280726570 + 239126570 + 2391267970 =

23912007 + 79702007 + 2807267970 + 239126

2858. Thirty Nine Countries by Frank Mrazik, Montreal, Quebec

Solve in base 16:

ALBANIA + ALGERIA + ANDORRA + ANGOLA + ARGENTINA + ARMENIA +

BAHAMAS + BAHRAIN + BANGLADESH + BARBADOS + BENIN + EAST + TIMOR +

ESTONIA + FINLAND + GABON + GAMBIA + GEORGIA + GHANA + GRENADA +

HAITI + INDIA + IRAN + ISRAEL + LEBANON + LESOTHO + LIBERIA + MALI +

MALTA + MONTENEGRO + NAMIBIA + NETHERLANDS + NIGER + NIGERIA +

ROMANIA + SAMOA + TOGO + TONGA + TRINIDAD + AND + TOBAGO =

   AFGHANISTAN

The only solution, where the standard hexadecimal representations are used for the digits over
10, a = 10, b = 11, c = 12, d= 13, e= 14 and f =15

9ce9729 + 9cdf129 + 9703119 + 97d3c9 + 91df75279 + 91af729 + e949a96 + e941927 +

e97dc90f64 + e91e9036 + ef727 + f965 + 52a31 + f653729 + 827c970 + d9e37 + d9ae29 +

df31d29 + d4979 + d1f7909 + 49252 + 27029 + 2197 + 2619fc + cfe9737 + cf63543 +

c2ef129 + a9c2 + a9c59 + a375f7fd13 + 79a2e29 + 7f54f1c9706 + 72df1 + 72df129 +

13a9729 + 69a39 + 53d3 + 537d9 + 51272090 + 970 + 53e9d3 = 98d49726597

2859. Skeleton – L by Junya Take, Kanagawa, Japan

```
      * * * * * *
        * * * * *
      _____

        * * * * * *
       * * L * * * *
        * * * L * * *
       * * * * L * *
        * * * * L *
      _____

      * * * * * L L L L * *
```

```
      461876
    ×  24891
  _____
      461876
    4156884
   3695008
  1847504
  923752
  _____

11496555516
```

2860. Skeleton – A by Junya Take, Kanagawa, Japan

```
      * * * * * * *
       * * * * *

      * * * * * * *
    * * A * * * * * *
    * * A * A * * * *
    * * * A * A * * *
    * * * * A A A *
    _____

* * * A * * * A * * * * *
```

```
      30722778
    × 209783
  _____
      92168334
     245782224
     215059446
     276505002
    61445556
  _____

  6445116537174
```

Proposers and solvers list for

Problems And Conjectures Appearing In 38(3)

P	S	Name	Location	28 50	28 51	28 52	28 53	28 54	28 55	28 56	28 57	28 58	28 59
*		Fred Barnes	Thayer. MO									P	
	*	Brian Barwell	Hampton, Middlesex, UK	S	S		S		S	S			S
	*	Lionel Bidwell	Cumbria, UK	S			S	S					
	*	Michael P. Cohen	Washington DC							S			
*		Andrew Cusumano	Great Neck, New York								P		
	*	Daniele Degiorgi	Massagno, Switzerland	S	S	S	S	S	S	S	S		S
*		Hubert Hagadorn	Meno Park, CA			P							
	*	Richard I. Hess	Rancho Palos Verdes, CA	S	S	S	S	S	S	S	S	S	S
*		Henry Ibstedt	Broby, Sweden		P		S		S		S		S
*		Robert Khanis	Sydney Australia	P									
*	*	Ken Klinger	Northbrook IL			S			P	S			
*		Donald E. Knuth	Stanford, CA				P						
*		Andy Pepperdine	Bath, UK						S				
*		Dario Uri	Pontecchio Marconi, Italy					P					
	*	Antoine Verroken		S					S				
*		John Wahl	Mt, Pocono, PA										P
*	*												
		Proposer/Solver	P = Proposer S = Solver										

Proposers and solvers list for

Problems And Conjectures Appearing In 38(4)

P	S	Name	Location	28	28	28	28	28	28	28	28	2	28
		Brian Barwell	Hampton, Middlesex, UK			S	S	S		S			S
		Lionel Bidwell	Cumbria, UK	S		S			S		S	S	
*		Phoebe Chua	Grantham, PA				P						
	*	Michael P. Cohen	Washington, DC				S			s			
	*	Daniele Degiorgi	Massagno, Switzerland	S		S	S	S	S	S	S	S	S
*	*	Hubert Hagadorn	Meno Park, CA	S	S		S		P			P	
	*	Richard I. Hess	Rancho Palos Verdes, CA	S		S	S		S	S	S	S	S
	*	Henry Ibstedt	Broby, Sweden	S		S	S		S	S	S	S	S
*		Donald E. Knuth	Stanford, CA	P									
	*	Kathleen Lewis	Brikama, the Gambia				S			S			
	*	Andy Pepperdine	Bath, UK				S			S			S
*		Frank Rubin	Wappinger Falls, New York		P								
*		John Runkle	Lima, Ohio							P			
*		Frank J. Swetz	Middletown, PA										P
*		Dario Uri	Pontecchio Marconi, Italy					P			P		
	*	Antoine Verroken											S
*	*	Proposer/Solver	P = Proposer S = Solver										

SOLUTIONS TO ALPHAMETICS APPEARING IN THIS ISSUE

by Charles Ashbacher

1.

```
      966
     2014
      105        and the 5 can be replaced by 7
    _____

     3085
```

2.

```
     6481
    13907
      462
    _____

    20950
```

3.

```
      2014
     13078        where 0 and 5 can interchange
     28539
     _____

     43631
```

4.

```
       33
     3948
     9870
   358868
   358868
   _____

   731587
```

5.

```
        1
     2749
       52        where 4 and 5 can interchange
     7071
     _____

     9873
```

6.

```
  62
 185
  23        where E can also be 0
 976

1246
```

7.

```
 307
4926
4926

10159
```

114

Errata

Henry Ibstedt pointed out that one of the biLLies on page 58 had only five squares rather than six. The structure of that biLLie should have been

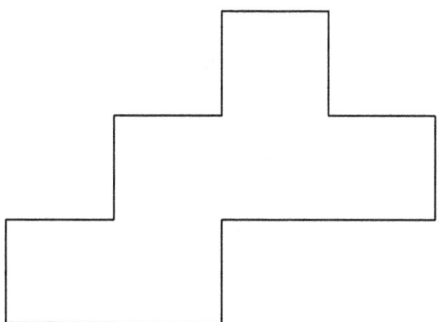

Symmetry—the aesthetics of mathematics *by Kate Jones*

Balance, equality, the yin and the yang—with the right combination of mosaic tilings we can achieve repeating or periodic patterns of great variety and beauty. The ancients, from India to Spain, recognized certain geometric groupings that would tile the infinite plane and identified 17 "symmetry groups", what we can call wallpaper patterns, composed of convex and concave polygons.

We started our little company, Kadon Enterprises, Inc., in 1979 to produce classic and original tiling sets. Some of them fit well on wallpaper grids, and all of them lend themselves to arrangements of surprisingly varied symmetries and startling beauty. We can find mirror (reflection) symmetry, rotational symmetry, opposite-color symmetry, rotations of opposite colors, congruent opposite colors, and symmetries in two-fold, three-fold, four- and even six-fold rotations. Here are a few of the puzzles we make, with a small sampling each of the amazing compositions they can produce. Feast your eyes!

Intarsia—16 each of just two kinds of tiles fill a hexagon in over 26 septillion ways and produce endless variations. *Top row:* opposite color rotational symmetry. *Center row:* rotational symmetry. *Bottom row, left:* mirror symmetry; *right,* opposite color reflection. Created by Henrik Morast, developed by Kate Jones.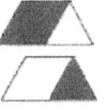

116

Chasing Squares—8 tiles each in two shapes and two colors, isosceles right triangles and "sheds", let you form inexhaustibly many designs. Invented by Jerry Farrell, developed by Kate Jones. See lots more patterns here: www.gamepuzzles.com/cs-designs.htm

All product names are proprietary trademarks of Kadon Enterprises, Inc.

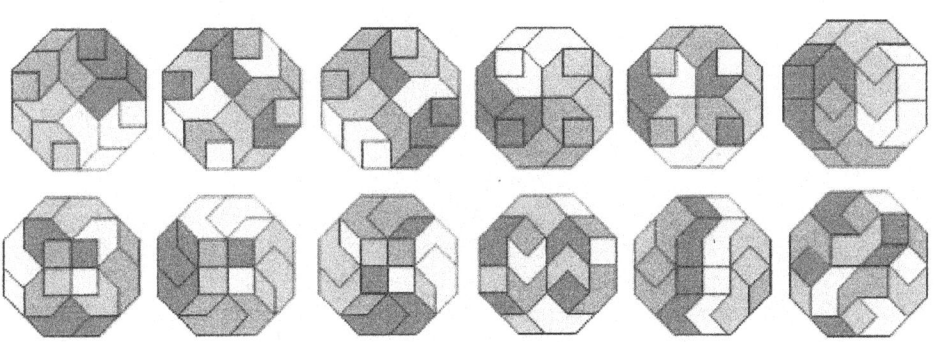

Rombix Jr.—4 shapes in 4 colors are based on Alan Schoen's rhombic dissection of an octagon and can form hundreds of colorful symmetry patterns. Great for ages 5 to adult.

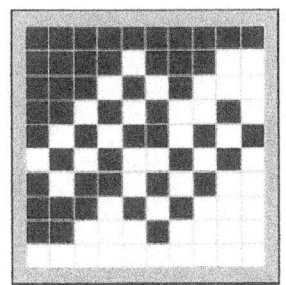 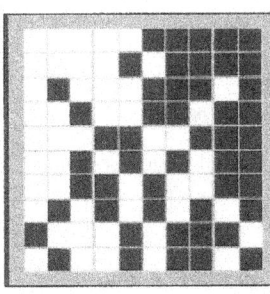  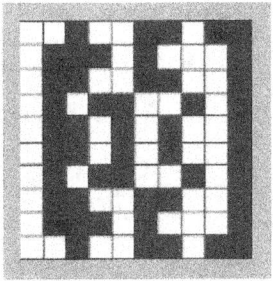

Quintapaths—20 sticks, each 1x5 differently inlaid with 0-5 black squares, form every kind of symmetry: rotational, mirror, opposite color. Created by Scott Kim, developed by Kate Jones.

All these puzzles are eloquent paradigms of how people think, solve, design, and organize systems. Out of diversity comesharmony. Most are suitable for ages 8 to adult. You can see these and more in the full website at **www.gamepuzzles.com**

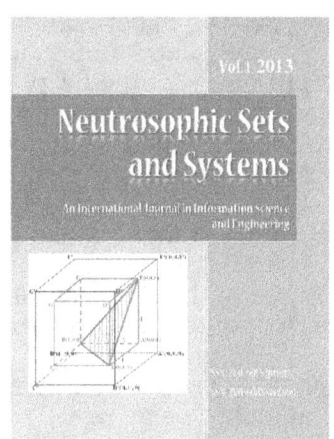

Editor-in-Chief:

Prof. Florentin Smarandache

Department of Mathematics and Science

University of New Mexico

705 Gurley Avenue

Gallup, NM 87301, USA

E-mail: smarand@unm.edu

Home page:
http://fs.gallup.unm.edu/NSS

Associate Editors:

Dmitri Rabounski and Larissa Borissova, independent researchers.

Said Broumi, Univ. of Hassan II Mohammedia, Casablanca, Morocco.

A. A. Salama, Faculty of Science, Port Said University, Egypt.

Yanhui Guo, School of Science, St. Thomas University, Miami, USA.

Francisco Gallego Lupiañez, Universidad Complutense, Madrid, Spain.

Peide Liu, Shandong Universituy of Finance and Economics, China.

Pabitra Kumar Maji, Math Department, K. N. University, WB, India.

S. A. Albolwi, King Abdulaziz Univ., Jeddah, Saudi Arabia.

Mohamed Eisa, Dept. of Computer Science, Port Said Univ., Egypt.

Neutrosophic Sets and Systems has been created for publications on advanced studies in neutrosophy, neutrosophic set, neutrosophic logic, neutrosophic probability, neutrosophic statistics that started in 1995 and their applications in any field, such as the neutrosophic structures developed in algebra, geometry, topology, etc.
  The submitted papers should be professional, in good English, containing a brief review of a problem and obtained results. Neutrosophy is a new branch of philosophy that studies the origin,

nature, and scope of neutralities, as well as their interactions with different ideational spectra.
 This theory considers every notion or idea <A> together with its opposite or negation <antiA> and with their spectrum of neutralities <neutA> in between them (i.e. notions or ideas supporting neither <A> nor <antiA>). The <neutA> and <antiA> ideas together are referred to as <nonA>.
 Neutrosophic Set and Logic are generalizations of the fuzzy set and respectively fuzzy logic (especially of intuitionistic fuzzy set and respectively intuitionistic fuzzy logic). In neutrosophic logic a proposition has a degree of truth (T), a degree of indeter
minacy (I), and a degree of falsity (F), where T, I, F are standard or non-standard subsets of $]^{-}0, 1^{+}[$.
 Neutrosophic Probability is a generalization of the classical probability and imprecise probability.
 Neutrosophic Statistics is a generalization of the classical statistics.
What distinguishes the neutrosophics from other fields is the <neutA>, which means neither <A> nor <antiA>.
 <neutA>, which of course depends on <A>, can be indeterminacy, neutrality, tie game, unknown, contradiction, ignorance, imprecision, etc.

All submissions should be designed in MS Word format using our template file:

   http://fs.gallup.unm.edu/NSS/NSS-paper-template.doc

A variety of scientific books in many languages can be downloaded freely from the Digital Library of Science:

   http://fs.gallup.unm.edu/eBooks-otherformats.htm

 To submit a paper, mail the file to the Editor-in-Chief. To order printed issues, contact the Editor-in-Chief. This journal is non-commercial, academic edition. It is printed from private donations.

Information about the neutrosophics you get from the UNM website:
   http://fs.gallup.unm.edu/neutrosophy.htm
The home page of the journal is accessed on
   http://fs.gallup.unm.edu/NSS

Topics in Recreational Mathematics
1/2015

Presenting papers and articles in recreational mathematics or material of interest to people interested in recreational mathematics. Original artwork with a mathematical theme will also be featured.

Contents

Editor-in-chief
Charles Ashbacher

Assistant editor
Rachel Pollari

Artwork
Caytie Ribble

Technical assistant
Gisela Hausmann

Dedicated to the legacy of Martin Gardner and Joseph S. Madachy

Available on Amazon

ISBN 978-1507603215

Topics in Recreational Mathematics
2/2015

Presenting papers and articles in recreational mathematics or material of interest to people interested in recreational mathematics. Original artwork with a mathematical theme will also be featured.

Editor-in-Chief: Charles Ashbacher
Assistant editor: Rachel Pollari
Artwork: Caytie Ribble
Technical advisor: Gisela Hausmann

Contents

ALPHAMETICS AS EXPRESSED IN RECREATIONAL MATHEMATICS MAGAZINE

Alphametics have been a staple of recreational mathematics since the first issue of **Recreational Mathematics Magazine**. A column of alphametics appeared in the first issue of RMM and it was a regular feature in Journal of Recreational Mathematics throughout the 38 ½ volumes that it was published.

This book contains the alphametics and their solutions that appeared in Recreational Mathematics Magazine during the 14 issues that it was published by Joseph S. Madachy.

Contents

Editor's Notes
by Charles Ashbacher

Mathematical Cartoon
by Caytie Ribble

Introduction
by Charles Ashbacher

Solving Addition Alphametics
by Charles Ashbacher

The Alphametics That Appeared in Recreational Mathematics Magazine

Solutions to Alphametics

Available on Amazon

ISBN 978-1508538134

Editor-in-chief
Charles Ashbacher

Artwork
Caytie Ribble

Technical assistant
Gisela Hausmann

TEN YEAR CUMULATIVE INDEX TO THE JOURNAL OF RECREATIONAL MATHEMATICS

Edited by

Joseph S. Madachy

Updated by Charles Ashbacher

This is a republication of the Ten Year Index published by Baywood Publishing Company in 1982.

Available on Amazon, ISBN 9781508936800

www.ingramcontent.com/pod-product-compliance
Lightning Source LLC
Chambersburg PA
CBHW080817180526
45168CB00006B/2478